Kinect in Motion – Audio and Visual Tracking by Example

A fast-paced, practical guide including examples, clear instructions, and details for building your own multimodal user interface

Clemente Giorio

Massimo Fascinari

BIRMINGHAM - MUMBAI

Kinect in Motion – Audio and Visual Tracking by Example

First published: April 2013

Production Reference: 1180413

Published by Packt Publishing Ltd.
Livery Place
35 Livery Street
Birmingham B3 2PB, UK.

ISBN 978-1-84969-718-7

www.packtpub.com

Cover Image by Suresh Mogre (suresh.mogre.99@gmail.com)

Credits

Authors

Clemente Giorio

Massimo Fascinari

Reviewers

Atul Gupta

Mandresh Shah

Acquisition Editor

James Jones

Commissioning Editor

Yogesh Dalvi

Technical Editors

Jalasha D'costa

Kirti Pujari

Project Coordinator

Sneha Modi

Proofreader

Paul Hindle

Indexer

Monica Ajmera Mehta

Production Coordinators

Pooja Chiplunkar

Nitesh Thakur

Cover Work

Pooja Chiplunkar

About the Authors

Clemente Giorio is an independent Consultant; he cooperated with Microsoft SrL for the development of a prototype that uses the Kinect sensor. He is interested in Human-computer Interface (HCI) and multimodal interaction.

I would first like to thank my family, for their continuous support throughout my time in University.

I would like to express my gratitude to the many people who saw me through this book. During the evolution of this book, I have accumulated many debts, only few of which I have space to acknowledge here.

Writing of this book has been a joint enterprise and a collaborative exercise. Apart from the names mentioned, there are many others who contributed. I appreciate their help and thank them for their support.

Massimo Fascinari is a Solution Architect at Avanade, where he designs and delivers software development solutions to companies throughout the UK and Ireland. His interest in Kinect and human-machine interaction started during his research on increasing the usability and adoption of collaboration solutions.

I would like to thank my wife Edyta, who has been supporting me while I was working on the book.

About the Reviewers

With more than 17 years of experience working on Microsoft technologies, **Atul Gupta** is currently a Principal Technology Architect at Infosys' Microsoft Technology Center, Infosys Labs. His expertise spans user experience and user interface technologies, and he is currently working on touch and gestural interfaces with technologies such as Windows 8, Windows Phone 8, and Kinect. He has prior experience in Windows Presentation Foundation (WPF), Silverlight, Windows 7, Deepzoom, Pivot, PixelSense, and Windows Phone 7.

He has co-authored the book *ASP.NET 4 Social Networking* (`http://www.packtpub.com/asp-net-4-social-networking/book`). Earlier in his career, he also worked on technologies such as COM, DCOM, C, VC++, ADO.NET, ASP.NET, AJAX, and ASP.NET MVC. He is a regular reviewer for Packt Publishing and has reviewed books on topics such as Silverlight, Generics, and Kinect.

He has authored papers for industry publications and websites, some of which are available on Infosys' Technology Showcase (`http://www.infosys.com/microsoft/resource-center/pages/technology-showcase.aspx`). Along with colleagues from Infosys, Atul blogs at `http://www.infosysblogs.com/microsoft`. Being actively involved in professional Microsoft online communities and developer forums, Atul has received Microsoft's Most Valuable Professional award for multiple years in a row.

Mandresh Shah is a developer and architect working in the Avanade group for Accenture Services. He has IT industry experience of over 14 years and has been predominantly working on Microsoft technologies. He has experience on all aspects of the software development lifecycle and is skilled in design, implementation, technical consulting, and application lifecycle management. He has designed and developed software for some of the leading private and public sector companies and has built industry experience in retail, insurance, and public services. With his technical expertise and managerial abilities, he also has played the role of growing capability and driving innovation within the organization.

Mandresh lives in Mumbai with his wife Minal, and two sons Veeransh and Veeshan. In his spare time he enjoys reading, movies, and playing with his kids.

www.PacktPub.com

Support files, eBooks, discount offers and more

You might want to visit www.PacktPub.com for support files and downloads related to your book.

Did you know that Packt offers eBook versions of every book published, with PDF and ePub files available? You can upgrade to the eBook version at www.PacktPub.com and as a print book customer, you are entitled to a discount on the eBook copy. Get in touch with us at service@packtpub.com for more details.

At www.PacktPub.com, you can also read a collection of free technical articles, sign up for a range of free newsletters and receive exclusive discounts and offers on Packt books and eBooks.

http://PacktLib.PacktPub.com

Do you need instant solutions to your IT questions? PacktLib is Packt's online digital book library. Here, you can access, read and search across Packt's entire library of books.

Why Subscribe?
- Fully searchable across every book published by Packt
- Copy and paste, print and bookmark content
- On demand and accessible via web browser

Free Access for Packt account holders

If you have an account with Packt at www.PacktPub.com, you can use this to access PacktLib today and view nine entirely free books. Simply use your login credentials for immediate access.

Table of Contents

Preface

To build interesting, interactive, and user friendly software applications, developers are turning to Kinect for Windows to leverage multimodal and Natural User Interface (NUI) capabilities in their programs.

Kinect in Motion – Audio and Visual Tracking by Example is a compact reference on how to master color, depth, skeleton, and audio data streams handled by Kinect for Windows. You will learn how to use Kinect for Windows for capturing and managing color images tracking user motions, gestures, and their voice.
This book, thanks to its focus on examples and to its simple approach, will guide you on how to easily step away from a mouse or keyboard driven application.

This will enable you to break through the modern application development space. The book will step you through many detailed, real-world examples, and even guide you on how to test your application.

What this book covers

Chapter 1, *Kinect for Windows – Hardware and SDK Overview*, introduces the Kinect, looking at the key architectural aspects such as the hardware composition and the software development kit components.

Chapter 2, *Starting with Image Streams*, shows you how to start building a Kinect project using Visual Studio and focuses on how to handle the color stream and the depth stream.

Chapter 3, *Skeletal Tracking*, explains how to track the skeletal data provided by the Kinect sensor and how to interpret them for designing relevant user actions.

Chapter 4, *Speech Recognition*, focuses on how to manage the Kinect sensor audio stream data and enhancing the Kinect sensor's capabilities for speech recognition.

Appendix, *Kinect Studio and Audio Recording*, introduces the Kinect Studio tool and shows you how to save and playback video and audio streams in order to simplify the coding and the test of our Kinect enabled application.

What you need for this book

The following hardware and software are required for the codes described in this book:

- CPU: Dual-core x86 or x64 at 2,66 Ghz or faster
- USB: 2.0 or compatible
- RAM: 2 GB or more
- Graphics card: DirectX 9.0c
- Sensor: Kinect for Windows
- Operating system: Windows 7 or Windows 8 (x86 and x64 version)
- IDE: Microsoft Visual Studio 2012 Express or an other edition
- Framework: .NET 4 or 4.5
- Software Development Kit: Kinect for Windows SDK
- Toolkit: Kinect for Windows Toolkit

The reader can also utilize a virtual machine (VM) environment from the following:

- Microsoft HyperV
- VMware
- Parallels

Who this book is for

This book is great for developers new to the Kinect for Windows SDK and those who are looking to get a good grounding in mastering the video and audio tracking. It's assumed that you will have some experience in C# and XAML already. Whether you are planning to use Kinect for Windows in your LOB application or for more consumer oriented software, we would like you to have fun with Kinect and to enjoy embracing a multimodal interface in your solution.

Conventions

In this book, you will find a number of styles of text that distinguish between different kinds of information. Here are some examples of these styles and an explanation of their meaning.

Code words in text are shown as follows: " The X8R8G8B8 format is a 32-bit RGB pixel format, in which 8 bits are reserved for each color."

A block of code is set as follows:

```
<Grid.RowDefinitions>
    <RowDefinition Height="Auto" />
    <!-- define additional RowDefinition entries as needed -->
</Grid.RowDefinitions>
```

When we wish to draw your attention to a particular part of a code block, the relevant lines or items are set in bold:

```
public partial class MainWindow : Window
{   private KinectSensor sensor;
    public MainWindow()
    {   InitializeComponent();
        this.Loaded += MainWindow_Loaded;
        KinectSensor.KinectSensors.StatusChanged += KinectSensors_
        StatusChanged;
    }
}
```

New terms and **important words** are shown in bold. Words that you see on the screen, in menus or dialog boxes for example, appear in the text like this: "Select the **WPF Application Visual C# template**".

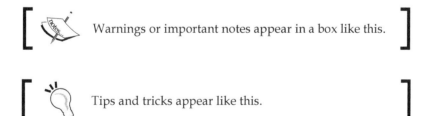

Warnings or important notes appear in a box like this.

Tips and tricks appear like this.

Reader feedback

Feedback from our readers is always welcome. Let us know what you think about this book—what you liked or may have disliked. Reader feedback is important for us to develop titles that you really get the most out of.

To send us general feedback, simply send an e-mail to feedback@packtpub.com and mention the book title in the subject of your message.

If there is a topic that you have expertise in and you are interested in either writing or contributing to a book, see our author guide on www.packtpub.com/authors.

Customer support

Now that you are the proud owner of a Packt book, we have a number of things to help you to get the most from your purchase.

Downloading the example code

You can download the example code files for all Packt books you have purchased from your account at http://www.packtpub.com. If you purchased this book elsewhere, you can visit http://www.packtpub.com/support and register to have the files e-mailed directly to you.

Downloading the color images of this book

We also provide you a PDF file that has color images of the screenshots/diagrams used in this book. The color images will help you better understand the changes in the output.

You can download this file from http://www.packtpub.com/sites/default/files/downloads/7187_Images.pdf.

Errata

Although we have taken every care to ensure the accuracy of our content, mistakes do happen. If you find a mistake in one of our books—maybe a mistake in the text or the code—we would be grateful if you would report this to us. By doing so, you can save other readers from frustration and help us improve subsequent versions of this book. If you find any errata, please report them by visiting http://www.packtpub. com/submit-errata, selecting your book, clicking on the **errata submission form** link, and entering the details of your errata. Once your errata are verified, your submission will be accepted and the errata will be uploaded on our website, or added to any list of existing errata, under the Errata section of that title. Any existing errata can be viewed by selecting your title from http://www.packtpub.com/support.

Piracy

Piracy of copyright material on the Internet is an ongoing problem across all media. At Packt, we take the protection of our copyright and licenses very seriously. If you come across any illegal copies of our works, in any form, on the Internet, please provide us with the location address or website name immediately so that we can pursue a remedy.

Please contact us at copyright@packtpub.com with a link to the suspected pirated material. We appreciate your help in protecting our authors, and our ability to bring you valuable content.

Questions

You can contact us at questions@packtpub.com if you are having a problem with any aspect of the book, and we will do our best to address it.

1
Kinect for Windows – Hardware and SDK Overview

In this chapter we will define the key notions and tips for the following topics:

- Critical hardware components of the Kinect for Windows device and their functionalities, properties, and limits
- Software architecture defining the Kinect SDK 1.6

Motion computing and Kinect

Before getting Kinect in motion, let's try to understand what motion computing (or motion control computing) is and how Kinect built its success in this area.

Motion control computing is the discipline that processes, digitalizes, and detects the position and/or velocity of people and objects in order to interact with software systems.

Motion control computing has been establishing itself as one of the most relevant techniques for designing and implementing a **Natural User Interface** (**NUI**).

NUIs are human-machine interfaces that enable the user to interact in a natural way with software systems. The goals of NUIs are to be natural and intuitive. NUIs are built on the following two main principles:

- The NUI has to be imperceptible, thanks to its intuitive characteristics: (a sensor able to capture our gestures, a microphone able to capture our voice, and a touch screen able to capture our hands' movements). All these interfaces are imperceptible to us because their use is intuitive. The interface is not distracting us from the core functionalities of our software system.

- The NUI is based on nature or natural elements. (the slide gesture, the touch, the body movements, the voice commands—all these actions are natural and not diverting from our normal behavior).

NUIs are becoming crucial for increasing and enhancing the user accessibility for software solution. Programming a NUI is very important nowadays and it will continue to evolve in the future.

Kinect embraces the NUIs principle and provides a powerful multimodal interface to the user. We can interact with complex software applications and/or video games simply by using our voice and our natural gestures. Kinect can detect our body position, velocity of our movements, and our voice commands. It can detect objects' position too.

Microsoft started to develop Kinect as a *secret project* in 2006 within the Xbox division as a competitive Wii killer. In 2008, Microsoft started Project Natal, named after the Microsoft General Manager of Incubation Alex Kipman's hometown in Brazil. The project's goal was to develop a device including depth recognition, motion tracking, facial recognition, and speech recognition based on the video recognition technology developed by PrimeSense.

Kinect for Xbox was launched in November 2010 and its launch was indeed a success: it was and it is still a break-through in the gaming world and it holds the Guinness World Record for being the "fastest selling consumer electronics device" ahead of the iPhone and the iPad.

In December 2010, PrimeSense (`primesense.com`) released a set of open source drivers and APIs for Kinect that enabled software developers to develop Windows applications using the Kinect sensor.

Finally, on June 17 2011 Microsoft launched the Kinect SDK beta, which is a set of libraries and APIs that enable us to design and develop software applications on Microsoft platforms using the Kinect sensor as a multimodal interface.

With the launch of the Kinect for Windows device and the Kinect SDK, motion control computing is now a discipline that we can shape in our **garages**, writing simple and powerful software applications ourselves.

This book is written for all of us who want to develop market-ready software applications using Kinect for Windows that can track audio and video and control motion based on NUI. In an area where Kinect established itself in such a short span of time, there is the need to consolidate all the technical resources and develop them in an appropriate way: this is our zero-to-hero Kinect in motion journey. This is what this book is about.

This book assumes that you have a basic knowledge of C# and that we all have a great passion to learn about programming for Kinect devices. This book can be enjoyed by anybody interested in knowing more about the device and learning how to track audio and video using the Kinect for Windows **Software Development Kit (SDK)** 1.6. We deeply believe this book will help you to master how to process video depth and audio stream and build market-ready applications that control motion. This book has deliberately been kept simple and concise, which will aid you to quickly grasp the core and critical concepts.

Before jumping on the core of audio and visual tracking with Kinect for Windows, let's take the space of this introduction chapter to understand what the hardware and software architectures Kinect for Windows and its SDK 1.6 use.

Hardware overview

The Kinect device is a horizontal bar composed of multiple sensors connected to a base with a motorized pivot.

The following image provides a schematic representation of all the main Kinect hardware components. Looking at the Kinect sensor from the front, from the outside it is possible to identify the Infrared (IR) Projector (1), the RGB camera (3), and the depth camera (2). An array of four microphones (6), the three-axis accelerometer (5), and the tilt motor (4) are arranged inside the plastic case.

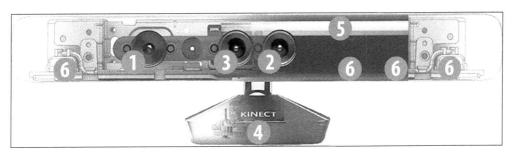

Kinect case and components

The device is connected to a PC through a USB 2.0 cable. It needs an external power supply in order to work because USB ports don't provide enough power.

Now let's jump in to the main features of its components.

The IR projector

The IR projector is the device that Kinect uses for projecting the IR rays that are used for computing the depth data. The IR projector, which from the outside looks like a common camera, is a laser emitter that constantly projects a pattern of structured IR dots at a wavelength around of 830 nm (patent US20100118123, Prime Sense Ltd.). This light beam is invisible to human eyes (that typically respond to wavelengths from about 390 nm to 750 nm) except for a red bright dot in the center of emitter.

The pattern is composed by 3 x 3 subpatterns of 211 x 165 dots (for a total of 633 x 495 dots). In each subpattern, one spot is much brighter than all the others.

As the dotted light (spot) hits an object, the pattern becomes distorted, and this distortion is analyzed by the depth camera in order to estimate the distance between the sensor and the object itself.

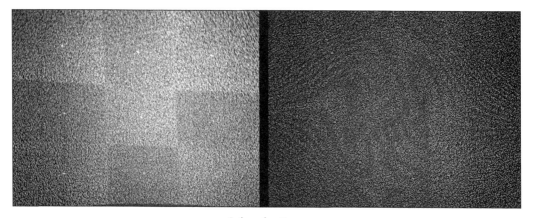

Infrared pattern

In the previous image, we tested the IR projector against the room's wall. In this case we have to notice that a view of the clear infrared pattern can be obtained only by using an external IR camera (the left-hand side of the previous image). Taking the same picture from the internal RGB camera, the pattern will look distorted even though in this case the beam is not hitting any object (the right-hand side of the previous picture).

Depth camera

The depth camera is a (traditional) monochrome CMOS (complementary metal-oxide-semiconductor) camera that is fitted with an IR-pass filter (which is blocking the visible light). The depth camera is the device that Kinect uses for capturing the depth data.

The depth camera is the sensor returning the 3D coordinates (x, y, z) of the scene as a stream. The sensor captures the structured light emitted by the IR projector and the light reflected from the objects inside the scene. All this data is converted in to a stream of frames. Every single frame is processed by the PrimeSense chip that produces an output stream of frames. The output resolution is upto 640 x 480 pixels. Each pixel, based on 11 bits, can represent 2048 levels of depth.

The following table lists the distance ranges:

Mode	Physical limits	Practical limits
Near	0.4 to 3 m (1.3 to 9.8 ft)	0.8 to 2.5 m (2.6 to 8.2 ft)
Normal	0.8 to 4 m (2.6 to 13.1 ft)	1.2 to 3.5 m (4 to 11.5 ft)

 The sensor doesn't work correctly within an environment affected by sunlight, a reflective surface, or an interference with light with a similar wavelength (830 nm circa).

The following figure is composed of two frames extracted from the depth image stream: the one on the left represents a scene without any interference. The one on the right is stressing how interference can reduce the quality of the scene. In this frame, we introduced an infrared source that is overlapping the Kinect's infrared pattern.

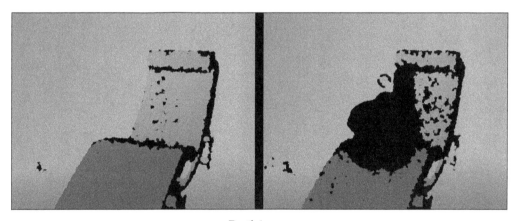

Depth images

The RGB camera

The RGB camera is similar to a common color webcam, but unlike a common webcam, the RGB camera hasn't got an IR-cut filter. Therefore in the RGB camera, the IR is reaching the CMOS. The camera allows a resolution upto 1280 x 960 pixels with 12 images per second speed. We can reach a frame rate of 30 images per second at a resolution of 640 x 480 with 8 bits per channel producing a Bayer filter output with a RGGBD pattern. This camera is also able to perform color flicker avoidance, color saturation operations, and automatic white balancing. This data is utilized to obtain the details of people and objects inside the scene.

The following monochromatic figure shows the infrared frame captured by the RGB camera:

IR frame from the RGB camera

To obtain high quality IR images we need to use dim lighting and to obtain high quality color image we need to use external light sources. So it is important that we balance both of these factors to optimize the use of the Kinect sensors.

Tilt motor and three-axis accelerometer

The Kinect cameras have a horizontal field of view of 57.5 degrees and a vertical field of view of 43.5 degrees. It is possible to increase the interaction space by adjusting the vertical tilt of the sensor by +27 and -27 degrees. The tilt motor can shift the Kinect head's angle upwards or downwards.

The Kinect also contains a three-axis accelerometer configured for a *2g* range (*g* is the acceleration value due to gravity) with a 1 to 3 degree accuracy. It is possible to know the orientation of the device with respect to gravity reading the accelerometer data.

The following figure shows how the field of view angle can be changed when the motor is tilted:

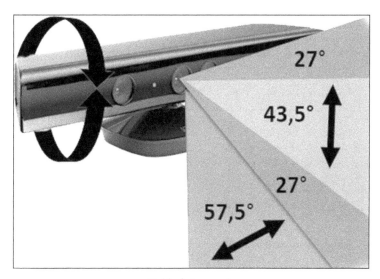

Field of view angle

Microphone array

The microphone array consists of four microphones that are located in a linear pattern in the bottom part of the device with a 24-bit **Analog to Digital Converter** (**ADC**). The captured audio is encoded using **Pulse Code Modulation** (**PCM**) with a sampling rate of 16 KHz and a 16-bit depth. The main advantages of this multi-microphones configuration is an enhanced Noise Suppression, an **Acoustic Echo Cancellation** (**AEC**), and the capability to determine the location and the direction of an audio source through a beam-forming technique.

Software architecture

In this paragraph we review the software architecture defining the SDK. The SDK is a composite set of software libraries and tools that can help us to use the Kinect-based natural input. The Kinect senses and reacts to real-world events such as audio and visual tracking. The Kinect and its software libraries interact with our application via the NUI libraries, as detailed in the following figure:

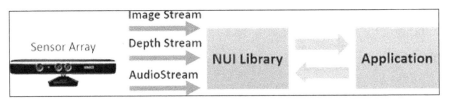

Interaction diagram

Here, we define the software architecture diagram where we encompass the structural elements and the interfaces by which the Kinect for Windows SDK 1.6 is composed, as well as the behavior as specified in collaboration with those elements:

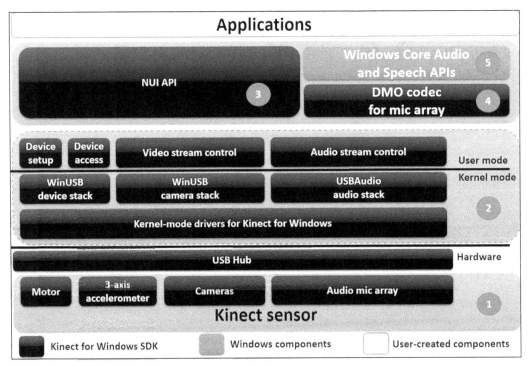

Kinect for Windows SDK 1.6 software architecture diagram

The following list provides the details for the information shown in the preceding figure:

- **Kinect sensor**: The hardware components as detailed in the previous paragraph, and the USB hub through which the Kinect sensor is connected to the computer.

- **Kinect drivers**: The Windows drivers for the Kinect, which are installed as part of the SDK setup process. The Kinect drivers are accessible in the `%Windows%\System32\DriverStore\FileRepository` directory and they include the following files:
  - `kinectaudio.inf_arch_uniqueGUID;`
  - `kinectaudioarray.inf_arch_uniqueGUID;`
  - `kinectcamera.inf_arch_uniqueGUID;`
  - `kinectdevice.inf_arch_uniqueGUID;`
  - `kinectsecurity.inf_arch_uniqueGUID`

  These files expose the information of every single Kinect's capabilities. The Kinect drivers support the following files:
  - The Kinect microphone array as a kernel-mode audio device that you can access through the standard audio APIs in Windows
  - Audio and video streaming controls for streaming audio and video (color, depth, and skeleton)
  - Device enumeration functions that enable an application to use more than one Kinect

- Audio and video components defined by NUI APIs for skeleton tracking, audio, and color and depth imaging. You can review the NUI APIs header files in the `%ProgramFiles%\Microsoft SDKs\Kinect\v1.6` folder as follows:
  - `NuiApi.h`: This aggregates all the NUI API headers
  - `NuiImageCamera.h`: This defines the APIs for the NUI image and camera services
  - `NuiSensor.h`: This contains the definitions for the interfaces as the `audiobeam`, the `audioarray`, and the accelerator
  - `NuiSkeleton.h`: This defines the APIs for the NUI skeleton

- **DirectX Media Object (DMO)** for microphone array beam-forming and audio source localization. The format of the data used in input and output by a stream in a DirectX DMO is defined by the `Microsoft.Kinect.DMO_MEDIA_TYPE` and the `Microsoft.Kinect.DMO_OUTPUT_DATA_BUFFER` structs. The default facade `Microsoft.Kinect.DmoAudioWrapper` creates a DMO object using a registered COM server, and calls native DirectX DMO layer directly.

- **Windows 7 standard APIs**: The audio, speech, and media APIs in Windows 7, as described in the Windows 7 SDK and the Microsoft Speech SDK (`Microsoft.Speech`, `System.Media`, and so on). These APIs are also available to desktop applications in Windows 8.

Video stream

The stream of color image data is handled by the `Microsoft.Kinect.ColorImageFrame`. A single frame is then composed of color image data. This data is available in different resolutions and formats. You may use only one resolution and one format at a time.

The following table lists all the available resolutions and formats managed by the `Microsoft.Kinect.ColorImageFormat` struct:

Color image format	Resolution	FPS	Data
InfraredResoluzion640x480Fps30	640 x 480	30	Pixel format is gray16
RawBayerResoluzion1280x960Fps12	1280 x 960	12	Bayer data
RawBayerResoluzion640x480Fps30	640 x 480	30	Bayer data
RawYuvResoluzion640x480Fps15	640 x 480	15	Raw YUV
RgbResoluzion1280x960Fps12	1280 x 960	12	RGB (X8R8G8B8)
RgbResoluzion640x480Fps15	640 x 480	15	Raw YUV
Undefined	N/A	N/A	N/A

When we use the `InfraredResoluzion640x480Fps30` format in the byte array returned for each frame, two bytes make up one single pixel value. The bytes are in little-endian order, so for the first pixel, the first byte is the least significant byte (with the least significant 6 bits of this byte always set to zero), and the second byte is the most significant byte.

The `X8R8G8B8` format is a 32-bit RGB pixel format, in which 8 bits are reserved for each color.

Raw YUV is a 16-bit pixel format. While using this format, we can notice the video data has a constant bit rate, because each frame is exactly the same size in bytes.

In case we need to increase the quality of the default conversion done by the SDK from Bayer to RGB, we can utilize the Bayer data provided by the Kinect and apply a customized conversion optimized for our central processing units (CPUs) or graphics processing units (GPUs).

> Due to the limited transfer rate of USB 2.0, in order to handle 30 FPS, the images captured by the sensor are compressed and converted in to RGB format. The conversion takes place before the image is processed by the Kinect runtime. This affects the quality of the images themselves.

In the SDK 1.6 we can customize the camera settings for optimizing and adapting the color camera for our environment (when we need to work in a low light or a brightly lit scenario, adapt contrast, and so on). To manage the code the `Microsoft.Kinect.ColorCameraSettings` class exposes all the settings we want to adjust and customize.

> In native code we have to use the `Microsoft.Kinect.Interop.INuiColorCameraSettings` interface instead.

In order to improve the external camera calibration we can use the IR stream to test the pattern observed from both the RGB and IR camera. This enables us to have a more accurate mapping of coordinates from one camera space to another.

Depth stream

The data provided by the depth stream is useful in motion control computing for tracking a person's motion as well as identifying background objects to ignore.

The depth stream is a stream of data where in each single frame the single pixel contains the distance (in millimeters) from the camera itself to the nearest object.

The depth data stream `Microsoft.Kinect.DepthImageStream` by the `Microsoft.Kinect.DepthImageFrame` exposes two distinct types of data:

- Depth data calculated in millimeters (exposed by the `Microsoft.Kinect.DepthImagePixel` struct).

- Player segmentation data. This data is exposed by the `Microsoft.Kinect.DepthImagePixel.PlayerIndex` property, identifying the unique player detected in the scene.

The following table defines the characteristics of the depth image frame:

Depth image format	Resolution	Frame rate
`Resoluzion640x480Fps30`	640 x 480	30 FPS
`Resoluzion320x240Fps30`	320 x 240	30 FPS
`Resolution80x60Fps`	80 x 60	30 FPS
`Undefined`	N/A	N/A

The Kinect runtime processes depth data to identify up to six human figures in a segmentation map. The segmentation map is a bitmap of `Microsoft.Kinect.DepthImagePixel`, where the `PlayerIndex` property identifies the closest person to the camera in the field-of-view. In order to obtain player segmentation data, we need to enable the skeletal stream tracking.

`Microsoft.Kinect.DepthImagePixel` has been introduced in the SDK 1.6 and defines what is called the "Extended Depth Data", or full depth information: each single pixel is represented by a 16-bit depth and a 16-bit player index.

 Note that the sensor is not capable of capturing infrared streams and color streams simultaneously. However, you can capture infrared and depth streams simultaneously.

Audio stream

Thanks to the microphone array, the Kinect provides an audio stream that we can control and manage in our application for audio tracking, voice recognition, high-quality audio capturing, and other interesting scenarios.

By default, Kinect tracks the loudest audio input. Having said that, we can certainly direct programmatically the microphone arrays (towards a given location, or following a tracked skeleton, and so on).

DirectX Media Object (DMO) is the building block used by Kinect for processing audio streams.

 In native scenario in addition to the DirectX Media Object (DMO), we can use the **Windows Audio Session API (WASAPI)** too.

In managed applications, the `Microsoft.Kinect.KinectAudioSource` class (exposed in the `KinectSensor.AudioSource` property) is the key software architecture component concerning the audio stream. Using the `Microsoft.Kinect.INativeAudioWrapper` class wraps the DirectX Media Object (DMO), which is a common Windows component for a single-channel microphone.

The `KinectAudioSource` class is not limited to wrap the DMO, but it introduces additional abilities such as:

- The `_MIC_ARRAY_MODE` as an additional microphone mode to support the Kinect microphone array.
- Beam-forming and source localization.
- The `_AEC_SYSTEM_MODE` **Acoustic Echo Cancellation (AEC)**. The SDK supports mono sound cancellation only.

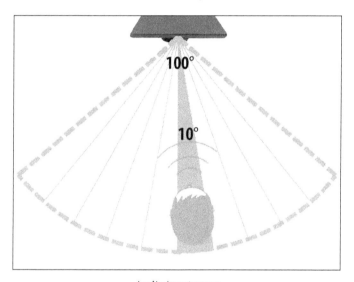

Audio input range

In order to increase the quality of the sound, audio inputs coming from the sensor get upto a 20 dB suppression. The array microphone allows an optional additional 6 dB of ambient noise removal for audio coming from behind the sensor.

The audio input has a range of +/– 50 degrees (as visualized in preceding figure) in front of the sensor. We can point the audio direction programmatically using a 10 degree increment range in order to focus our attention on a given user or to elude noise sources.

Skeleton

In addition to the data provided by the depth stream, we can use those provided by the skeleton tracking to enhance the motion control computing capabilities of our applications in regards to recognizing people and following their actions.

We define the *skeleton* as a set of *positioned key points*. A detailed skeleton contains 20 points in normal mode and 10 points in seated mode, as shown in the following figure. Every single point of the skeleton highlights a joint of the human body.

Thanks to the depth (IR) camera, Kinect can recognize up to six people in the field of view. Of these, up to two can be tracked in detail.

The stream of skeleton data is maintained by the `Microsoft.Kinect.SkeletonStream` class and the `Microsoft.Kinect.SkeletonFrame` class. The skeleton data is exposed for each single point in the 3D space by the `Microsoft.Kinect.SkeletonPoint` struct. In any single frame handled by the skeleton stream we can manage up to six skeletons using an array of the `Microsoft.Kinect.Skeleton` class.

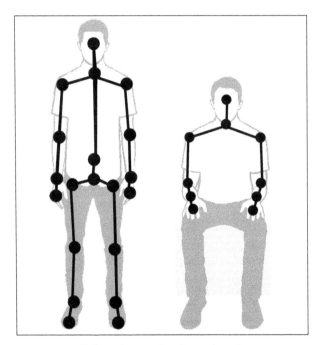

Skeleton in normal and seated mode

Summary

In this chapter we introduced Kinect, looking at the key architectural aspects such as the hardware composition and the SDK 1.6 software components. We walked through the color sensor, IR depth sensors, IR emitter, microphone arrays, the tilt motor for changing the Kinect camera angles, and the three-axis accelerometer.

Kinect generates two video streams using the color camera data and the depth information using the depth sensor. Kinect can detect up to six users in its view field and produce a detailed skeleton for two of them. All these characteristics make Kinect an awesome tool for video tracking motion. The Kinect's audio tracking makes the device a remarkable interface for voice recognition. Combining video and audio, Kinect and its SDK 1.6 are an outstanding technology for NUI.

Kinect is not just technology, it is indeed a means of how we can elevate the way users interact with complex software applications and systems. It is a break-through on how we can include NUIs and multimodal interface.

Kinect discloses unlimited opportunities to developers and software architects to design and create modern applications for different industries and lines of business.

The following examples are not meant to be an exhaustive list, but just a starting point that can inspire your creativity and increase your *appetite* for this technology.

- **Healthcare**: This improves the physical rehabilitation process by constantly capturing data of the motion and posture of patient. We can enhance this scenario by allowing doctors to check the patient data remotely streamed by the Kinect sensor.

- **Education/Professional development**: This helps in creating safe and more engaging environments based on gamification where students, teachers, and professionals can exercise activities and knowledge. The level of engagement can be increased even further using augmented reality.

- **Retail**: This engages customers across multiple channels using the Kinect's multimodal interface. Kinect can be used as a navigation system for virtual windows while shopping online and/or visiting infotainment kiosks.

- **Home automation**: This is also known as domotics where, thanks to the Kinect audio and video tracking, we can interact with all the electrical devices installed at our home (lights, washing machine, and so on).

In the next chapter, we will start to develop with the Kinect SDK, utilizing the depth and RGB camera streams. The applied examples will enable our application to optimize the way we manage and tune the streams themselves.

2
Starting with Image Streams

The aim of this chapter is to understand the steps for capturing data from the color stream, depth stream, and IR stream data. The key learning tools and steps for mastering all these streams are:

- **color camera**: data stream, event driven and polling techniques to manage color frames, image editing, color image tuning, and color image formats

- **depth image**: data stream, depth image ranges, and mapping between color image and depth image

All the examples we will develop in this book are built on Visual Studio 2010 or 2012. In this introduction, we want to include the key steps for getting started.

From Visual Studio, select **File | New | Project**. In the **New Project** window, do the following:

1. Select the **WPF Application Visual C#** template.
2. Select the **.Net Framework 4.0** as the framework for the project (it works in .Net Framework 4.5 too).
3. Assign a name to the project (in our example, we selected Chapter02)
4. Choose a location for the project.
5. Leave all the other settings with the default value.
6. Click on the **OK** button.

In the **Solution Explorer** window, please locate the references of the project. Right-click on **References** and select **Add Reference** to invoke the **Reference Manager** window. Select the **Microsoft.Kinect Version 1.6.0.0** assembly and click on the **OK** button.

An alternative approach to speeding up the preceding steps is to consider downloading the KinectContrib (`http://kinectcontrib.codeplex.com`) Visual Studio templates.

Color stream

Let's start by focusing on the color stream data. We are going to develop an example of how to apply data manipulation to the captured color stream.

The complete code is included in the `CODE_02/ColorStream` example delivered together with this book.

In the `MainWindows.xaml` file defined in the Visual Studio project, let's design our User Interface (UI) elements. We will use those elements to display the data obtained from the color stream.

Within the `<Grid>  </Grid>` tags we can add the following XAML code:

```
<Grid.RowDefinitions>
    <RowDefinition Height="Auto" />
    <!-- define additional RowDefinition entries as needed -->
</Grid.RowDefinitions>
<Grid.ColumnDefinitions>
    <ColumnDefinition Width="Auto" />
    <!-- define additional ColumnDefinition entries as needed -->
</Grid.ColumnDefinitions>
<Image Name="imgMain" Grid.Row="1" Grid.Column="1" Grid.ColumnSpan="2"
/>
<TextBlock Name="tbStatus" Grid.Row="3" Grid.Column="2" />
```

The `<Grid.RowDefinitions>` and `<Grid.ColumnDefinitions>` tags define the UI layout and the set of placeholders for additional UI elements, which we will use later in the example. The `imgMain` image is the control we will use to display the color stream data and the `tbStatus TextBlock` is the control we will use for providing the feedback on the Kinect sensor status.

To get our color data displayed we need to first of all initialize the sensor. Here are the tasks for initializing the sensor to generate color data.

In the `MainWindows.xaml.cs` file we enhance the code generated by Visual Studio by performing the following steps:

- Retrieving the available sensors and selecting the first one (if any) connected at any time using the `private KinectSensor sensor` member

- Enabling the color stream using the `KinectSensor.ColorStream.Enable(ColorImageFormat colorImageFormat)` API

- Starting the Kinect sensor using the `KinectSensor.Start()` API

Our code will look like this:

```
public partial class MainWindow : Window
{    private KinectSensor sensor;

    public MainWindow()
    {    InitializeComponent();
        this.Loaded += MainWindow_Loaded;
        KinectSensor.KinectSensors.StatusChanged += KinectSensors_
StatusChanged;
    }

    void MainWindow_Loaded(object sender, RoutedEventArgs e)
    {    this.tbStatus.Text = Properties.Resources.KinectInitialising;
        //Invoke the Kinect Status changed at the app start-up
        KinectSensors_StatusChanged(null, null);
    }

//handle the status changed event for the current sensor.
    void KinectSensors_StatusChanged(object
sender,StatusChangedEventArgs e)
  {   //select the first (if any available) connected Kinect Sensor
      from the KinectSensor.KinectSensors collection
 this.sensor = KinectSensor.KinectSensors.FirstOrDefault(s => s.Status
== KinectStatus.Connected);
    if (null != this.sensor)
 {this.tbStatus.Text = Properties.Resources.KinectReady;
//Color Image initialised
InitializeColorImage(ColorImageFormat.RgbResolution640x480Fps30);
            // Start the sensor
            try { this.sensor.Start();}
            catch (IOException) {this.sensor = null;}
  }
```

```
if (null == this.sensor)
        {this.tbStatus.Text = Properties.Resources.NoKinectReady;}
}

   void InitializeColorImage(ColorImageFormat colorImageFormat)
   { // Turn on the color stream to receive color frames
       this.sensor.ColorStream.Enable(colorImageFormat);}

   }
```

In order to compile the previous code we need to resolve the `Microsoft.Kinect` and `System.IO` namespaces. The values assigned to the `tbStatus.Text` are defined as `Properties` in the `Resources.resx` file.

The `KinectSensor.KinectColorFrameReady` event is the event that the sensor fires when a new frame from the color stream data is ready. The Kinect sensor streams out data continuously, one frame at a time, till we enforce the sensor to stop — using `KinectSensor.Stop()` — or we disable the color stream itself — using `KinectSensor.ColorStream.Disable()`.

We can register to this event to process the color stream data available and implement the related event handler.

After the `InitializeColorImage` method call, let's add the `ColorFrameReady` event of the `Sensor` object to process the color stream data. We manage the event defining the following event handler:

```
private void SensorColorFrameReady(object sender,
ColorImageFrameReadyEventArgs e)
{
    using (ColorImageFrame colorFrame = e.OpenColorImageFrame())
    { if (colorFrame != null) {   } }
}
```

By testing `colorFrame != null`, we ensure that the `colorFrame` object has been rendered smoothly.

We now need to copy the data to our local memory in order to manipulate the same and make them available for the `imgMain` image control defined in the UI.

In the `MainWindow` class, we define the `private byte[] colorPixels` variable, which is going to store the data received by the color stream.

We need to pre-allocate the byte array, `colorPixels`, for containing all the pixels stored in the color frame and provided by the `Int32 Image.PixelDataLength` property.

We need to define the `private WriteableBitmap colorBitmap` instance to hold the color information obtained by the color stream data.

Our `InitializeColorImage` method will now look like:

```
void InitializeColorImage(ColorImageFormat colorImageFormat)
{    // Turn on the color stream to receive color frames
    this.sensor.ColorStream.Enable(colorImageFormat);

    //Allocate the array to contain pixels stored in the color frame
    this.colorPixels = new byte[this.sensor.ColorStream.
FramePixelDataLength];

    //Create the WriteableBitmap with the appropriate PixelFormats
    this.colorBitmap = new WriteableBitmap(this.sensor.ColorStream.
FrameWidth,
            this.sensor.ColorStream.FrameHeight, 96.0, 96.0,
PixelFormats.Bgr32, null);
    // Displaying to point to the bitmap where putting image data
    this.imgMain.Source = this.colorBitmap;
}
```

Using the `ColorImageFrame.CopyPixelDataTo(byte[] colorPixels)` API, we copy the color frame's pixel data to our pre-allocated byte array `colorPixels`.

Finally, after getting the color data and saving it in the `WriteableBitmap` object, we draw the `WriteableBitmap` object itself using the `WritePixels(Int32Rect sourceRect, Array pixels, int stride, int offset)` method. In computing the `stride` parameter we have to take into account the `BytesPerPixel` value of the `ColorImageFrame` in relation to the `ColorImageFormat`. In this current example, as we are dealing with an **RGBA** (Red Green Blue Alpha) `ColorImageFormat`, the `BytesPerPixel` value is 4.

Let's now complete the body of the `if (colorFrame != null)` selection introduced previously in the event handler:

```
if (colorFrame != null)
{    //copy the color frame's pixel data to the array
    colorFrame.CopyPixelDataTo(this.colorPixels);

    //draw the WritableBitmap
    this.colorBitmap.WritePixels(
```

```
            new Int32Rect(0, 0, this.colorBitmap.PixelWidth, this.
  colorBitmap.PixelHeight),
            this.colorPixels, this.colorBitmap.PixelWidth *
  colorFrame.BytesPerPixel, 0);
  }
```

Compiling and running our example in Visual Studio, we are now in business.

Since our `SensorColorFrameReady` method runs frequently, this code maximizes performance by doing the minimum processing necessary to get the new data and copy it to the local memory. How do we improve performance?

The `using` statement automatically takes care of disposing of the `ColorImageFrame` object when we are done using it.

allocating the memory for the byte array `colorPixels` outside the event handler.

Using the `WriteableBitmap` array instead of creating a Bitmap for every frame. We can create the `WriteableBitmap` array only when the pixel format changes.

Editing the colored image

We can now think about manipulating the color stream data and applying some effects to enhance our example output. The following code provides a compact sample of how we can add a sphere effect to the left half of the image:

```
private void Sphere(int width, int height)
{   int xMid = width / 2; int yMid = height / 2;
  for (int x = 0; x < xMid; x++)
    {   for (int y = 0; y < height; ++y)
        {   //Compute the angle between the real point vs the center
        int trueX = x - xMid; int trueY = y - yMid;
        var theta = Math.Atan2(trueY, trueX);
        double radius = Math.Sqrt(trueX * trueX + trueY * trueY);
        double newRadius = radius * radius / (Math.Max(xMid, yMid));
        //Compute the distortion as projection of the new angle
        int newX = Math.Max(0, Math.Min((int)(xMid + (newRadius *
        Math.Cos(theta))), width - 1));
        int newY = Math.Max(0, Math.Min((int)(yMid + (newRadius *
        Math.Sin(theta))), height - 1));
        int pOffset = ((y * width) + x) * 4;
        int newPOffset = ((newY * width) + newX) * 4;
```

```
//draw the new point
colorPixels[pOffset] = colorPixels[newPOffset];
colorPixels[pOffset + 1] = colorPixels[newPOffset + 1];
colorPixels[pOffset + 2] = colorPixels[newPOffset + 2];
colorPixels[pOffset + 3] = colorPixels[newPOffset + 3];
}
}}
```

 Please note that the example included with this book provides a full sphere effect. The example includes additional sample algorithms to apply effects as: Pixelate, Flip, and RandomJitter.

The image manipulation or effects need to be applied just before the `this.colorBitmap.WritePixels` call within the `SensorColorFrameReady` event handler. As stated previously, this event handler runs frequently, so we need to ensure that its execution is performing.

What if the next `KinectColorFrameReady` event is fired before the image manipulation has completed?

This is a very likely scenario, as the Kinect sensor streams data with a throughput of 40 to 60 milliseconds circa and the image manipulations are usually heavy and long processing activities.

In this case, we have to change the technique by which we process the color stream data and apply instead what is called the **polling** approach.

In the polling approach we don't obtain the frame of the color stream data subscribing to the `KinectSensor.KinectColorFrameReady` event, but we request a new frame enquiring the `ColorImageFrame OpenNextFrame (int millisecondsWait)` API exposed by the `KinectSensor.ColorStream` object.

To implement this scenario, first of all we need to create a `BackgroundWorker` class instance that is able to run the color frame handling asynchronously, and update the `WritableBitmap` on the UI thread:

Subscribing to the `BackgroundWorker.DoWork` event, we ensure that intensive manipulation of the color frame is performed asynchronously leaving the UI thread free to respond to all the user inputs.

```
void backgroundWorker1_DoWork(object sender, DoWorkEventArgs e)
{    using (ColorImageFrame colorFrame = this.sensor.ColorStream.
    OpenNextFrame(0))
    {    if (colorFrame != null)
        {    //copy the color frame's pixel data to the array
            colorFrame.CopyPixelDataTo(this.colorPixels);
            //potentially apply the image effect
```

```
if (this.dataContext.IsImageEffectApplied)
{    Sphere(colorFrame.Width, colorFrame.Height,
     colorFrame.BytesPerPixel);
     Pixelate(colorFrame.Width, colorFrame.Height,
     20,colorFrame.BytesPerPixel);
}//update the approximate value of the frame rate
UpdateFrameRate();   }}}
```

Once the color frame manipulation is completed, we keep invoking the
`BackgroundWorker.RunWorkerAsync()` method in order to continuously
capture the next frame.

The first time the `BackgroundWorker.RunWorkerAsync()` method is invoked, just
after the `this.sensor.Start();`, call in the `KinectSensors_StatusChanged` method:

```
private void backgroundWorker1_RunWorkerCompleted(
    object sender, RunWorkerCompletedEventArgs e){
    this.colorBitmap.WritePixels(
        new Int32Rect(0, 0, this.colorBitmap.PixelWidth, this.
        colorBitmap.PixelHeight),
            this.colorPixels,
            this.colorBitmap.PixelWidth * 4, 0);
    //restart the manipulation
    this.backgroundWorker1.RunWorkerAsync();}
```

Using the new asynchronous pattern, we can take advantage of the
`async` and `await` keywords and simplify the previous syntax.

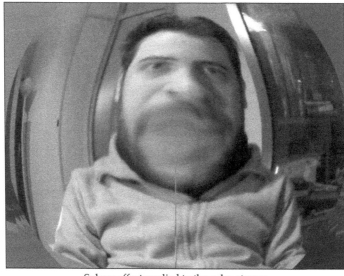

Sphere effect applied to the color stream

Image tuning

In order to improve or tweak the stream image quality, the SDK 1.6 provides color camera settings that allows you to add basic filters and effects on frames without the need to implement a customized one. With these settings and the processing of raw sensor data, it is possible to balance the different kinds of environmental light conditions.

This is done simply by using the `KinectSensor.ColorStream.CameraSettings` class, which implements the `InotifyPropertyChanged` interface. We can tune the camera to collect more useful color image data utilizing the properties of the following table:

Properties	Possible values	Default
`BacklightCompensationMode` gets or sets the compensation mode for adjusting the camera to capture data dependent on the environment	`AverageBrightness`, `CenterOnly`, `CenterPriority`, `LowlightsPriority`	`Average Brightness`
`Brightness` gets or sets the brightness or lightness	Range: [0.0, 1.0]	0.2156
`Contrast` gets or sets the amount of difference between lights and darks	Range: [0.5, 2.0]	1.0
`ExposureTime` gets or sets the exposure time	Range: [0.0, 4000] Increments: 1/10,000	0.0
`FrameInterval` gets or sets the frame interval	Range: [0, 4000] Increments: 1/10,000	0
`Gain` gets or sets the multiplier for the RGB color values	Range: [1.0, 16.0]	1.0
`Gamma` gets or sets the nonlinear operations for coding luminance data	Range: [1.0, 2.8]	2.2
`Hue` gets or sets the value that describes the shade of a color	Range: [-22.0, 22.0]	0.0
`PowerLineFrequency` reduces the flicker caused by the frequency of a power line	Disabled, FittyHertz, SixtyHertz	Disabled
`Saturation` gets or sets the colorfulness of a color relative to its own brightness	Range: [0.0, 2.0]	1.0
`Sharpness` Gets or sets the amount of detail visible.	Range: [0.0, 1.0]	0.5
`WhiteBalance` gets or sets the color temperature in degrees Kelvin	Range: [2700, 6500]	2700

The CameraSettings class has the AutoExposure and AutoWhiteBalance properties, which are by default set to true. In order to change the Exposure and WhiteBalance settings it is necessary to set those properties to false before using the color frame.

Use the ResetToDefault() method in order to reset the color camera setting to its default value.

The color image formats

The Bayer formats match the physiology of the human eye better than other image formats because they include more green pixels values than blue or red.

The Bayer color image data that the Kinect sensor returns at 1280 x 960 is actually compressed and converted to RGB before transmission to the runtime. The runtime then decompresses the data before it passes the data to the application. The compression is performed in order to return color data at frame rates as high as 30 fps. Naturally this has the side effect of some loss in the image fidelity.

Be aware that Direct3D and OpenGL use different byte orders. Direct3D (and Kinect for Windows) color data uses ARGB order, where alpha is in the highest-order byte and blue is in the lowest-order byte. OpenGL color data uses BGRA order, where blue is the highest-order byte and alpha is the lowest-order byte. To convert color data between the two formats, you will need to shift from one component order to the other to get the same results. The complete details are provided at http://msdn.microsoft.com/en-us/library/jj131027.aspx

Additional information on Bayer formats and filters is available at http://en.wikipedia.org/wiki/Bayer_filter.

The API provided by the SDK V 1.6 allows us to change dynamically the format of the color stream data. In order to change the format of the color image we can set the ColorImageFrame.Format property with the appropriate ColorImageFormat value.

Changing the ColorImageFrame.Format property in our source code is the only update we need to implement when using the following enum values:

- ColorImageFormat.RgbResolution1280x960Fps12:

  (RGB format—1280 x 960 pixels as size—12 frames per second)

- ColorImageFormat.RgbResolution640x480Fps30:

  (RGB format—640 x 480 pixels as size—30 frames per second)

- `ColorImageFormat.YuvResolution640x480Fps15:`

  (YUV format — 640 x 480 pixels as size — 15 frames per second)

For all the other formats, we need to take into account additional considerations and steps.

The Infrared color image format

The Kinect sensor streams out the IR data using the same channel of the color stream data. For us to display the IR data stream data, we need to perform the following steps:

1. Set the `ColorImageFrame.Format` property to the `ColorImageFormat.InfraredResolution640x480Fps30` value.

2. Ensure that `colorBitmap` is initialized with the `PixelFormats.Gray16` format in order to allocate the right amount of bytes, as shown in the following code snippet:

```
//Create the WriteableBitmap with the appropriate PixelFormats
if (colorImageFormat == ColorImageFormat.
InfraredResolution640x480Fps30)
{   //16 bits-per-pixel grayscale channel, 65536 shades of gray
    this.colorBitmap = new WriteableBitmap(this.sensor.
ColorStream.FrameWidth,
          this.sensor.ColorStream.FrameHeight, 96.0, 96.0,
PixelFormats.Gray16, null); }
```

> Setting the `ColorImageFrame.Format` property to `ColorImageFormat.InfraredResolution640x480Fps30` enables the Kinect sensor to work in Infrared IR mode.
>
> In low-light situations or without any environmental light the only way to obtain video data stream is to use the IR mode.

The raw Bayer formats

The Kinect sensor is able to provide the color stream data in raw Bayer format using the two color image formats: `RawBayerResolution1280x960Fps12` and `RawBayerResolution640x480Fps30`.

To display the color stream based on the two raw Bayer formats, we need to:

- Set the `ColorImageFrame.Format` to `ColorImageFormat.RawBayerResolution1280x960Fps12` or `ColorImageFormat.RawBayerResolution640x480Fps30`.

- In addition to the intermediate storage, `colorPixels` defines the color data as an array of bytes, while the `rawColorPixels` defines the intermediate storage for the color raw data. These intermediate storages will be used in the conversion from Bayer format to RGB format.

- The two storages need to be sized as:

```
this.rawColorPixels = new byte[colorFrame.PixelDataLength];
//byetsPerPixel in this case is 4
this.colorPixels = new byte[bytesPerPixel * colorFrame.Width *
colorFrame.Height];
```

- Copy the pixels contained in the color frame to the intermediate storage for the color raw data as follows:

```
colorFrame.CopyPixelDataTo(this.rawColorPixels);
```

- Convert the data contained in `rawColorPixels` to `colorPixels`.

- We are now ready to display the raw Bayer color data in the image control.

A valid Bayer to RGB conversion algorithm is defined by the following code snippet:

```
private  void ConvertBayerToRgb32(int width, int height)
{
  // Demosaic using a basic nearest-neighbor algorithm
    for (int y = 0; y < height; y += 2)
    {
      for (int x = 0; x < width; x += 2)
        {
            int firstRowOffset = (y * width) + x;
            int secondRowOffset = firstRowOffset + width;

            // Cache the Bayer component values.
            byte red = rawColorPixels[firstRowOffset + 1];
            byte green1 = rawColorPixels[firstRowOffset];
            byte green2 = rawColorPixels[secondRowOffset + 1];
            byte blue = rawColorPixels[secondRowOffset];

            // Adjust offsets for RGB.
            firstRowOffset *= 4; secondRowOffset *= 4;
            // Top left
```

```
        colorPixels[firstRowOffset] = blue;
        colorPixels[firstRowOffset + 1] = green1;
        colorPixels[firstRowOffset + 2] = red;
        // Top right
        colorPixels[firstRowOffset + 4] = blue;
        colorPixels[firstRowOffset + 5] = green1;
        colorPixels[firstRowOffset + 6] = red;
        // Bottom left
        colorPixels[secondRowOffset] = blue;
        colorPixels[secondRowOffset + 1] = green2;
        colorPixels[secondRowOffset + 2] = red;
        // Bottom right
        colorPixels[secondRowOffset + 4] = blue;
        colorPixels[secondRowOffset + 5] = green2;
        colorPixels[secondRowOffset + 6] = red;
}}}
```

Warning

We may hit an argument exception stating that **Buffer size is not sufficient** if the right sizing of the rawColorPixes and colorPixes are not implemented and the conversion is not performed. Please also provide the reasoning of why this exception would occur.

YUV raw format

The Kinect sensor is able to provide the color stream data in YUV raw format. The Kinect sensor adopts the YUV 4:2:2 standards, which use 4 bytes per 2 pixels. We can set the YUV raw format using the ColorImageFormat. RawYuvResolution640x480Fps15.

For displaying a color stream based on the YUV raw format, we perform similar steps to those described for the raw Bayer format. The key differences are:

- Set the ColorImageFrame.Format to ColorImageFormat. RawYuvResolution640x480Fps15

- Convert the data contained in rawColorPixels to colorPixels

A valid YUV 4:2:2 to RGB conversion algorithm is defined by the following code snippet:

```
private void ConvertYuvToRgb32(int width, int height)
{   for (int i = 0; i < rawColorPixels.Length- 4; i ++)
    {          // Cache the YUV component values.
            byte Y1 = rawColorPixels[i + 1];
```

```
                byte U = rawColorPixels[i];
                byte V = rawColorPixels[i + 2];
                byte Y2 = rawColorPixels[i + 3];
                int C1 = Y1 - 16; int D = U - 128;
                int E = V - 128; int C2 = Y2 - 16;
//Apply the YUV444 to RGB conversion we need to convert to byte
  colorPixels[i + 2] = ((298 * C1 + 409 * E + 128) >> 8);
  colorPixels[i + 1] = ((298 * C1 - 100 * D - 208 * E + 128) >> 8);
  colorPixels[i] = ((298 * C1 + 516 * D + 128) >> 8);
  colorPixels[i + 6] = ((298 * C2 + 409 * E + 128) >> 8);
  colorPixels[i + 5] = ((298 * C2 - 100 * D - 208 * E + 128) >> 8);
  colorPixels[i + 4] = ((298 * C2 + 516 * D + 128) >> 8);      }}
```

The YUV to RGB conversion algorithm applied in this chapter is based on the following formulas for the conversion between the two formats:

RGB to YUV conversion

$Y = (0.257 * R) + (0.504 * G) + (0.098 * B) + 16$

$Cr = V = (0.439 * R) - (0.368 * G) - (0.071 * B) + 128$

$Cb = U = -(0.148 * R) - (0.291 * G) + (0.439 * B) + 128$

YUV to RGB conversion

$B = 1.164(Y - 16) + 2.018(U - 128)$

$G = 1.164(Y - 16) - 0.813(V - 128) - 0.391(U - 128)$

$R = 1.164(Y - 16) + 1.596(V - 128)$

The source code project attached to this paragraph covers all the aspects and considerations developed in the paragraph itself. The `CODE_02/ColorStream` example covers the color stream data manipulation, the polling technique, and all the color image formats.

The `CODE_02/Tuning` example covers the techniques for adjusting the image by the SDK V 1.6 camera setting's APIs.

Depth stream

The process, and the relating code for getting our depth stream data displayed, is very similar to the one we detailed for the color stream data.

In this section, we will list and document the essential steps for working with the depth stream data. The example attached to this chapter will provide all the additional details.

In order to process the depth stream data obtained by the connected `KinectSensor, sensor` we need to enable the `KinectSensor.DepthStream` using the `KinectSensor.DepthStream.Enable(ColorImageFormat colorImageFormat)` API.

The `KinectSensor.DepthFrameReady` is the event that the sensor fires when a new frame from the depth stream data is ready. The Kinect sensor streams data out continuously, one frame at a time, till the sensor is stopped or the depth stream is disabled. To stop the sensor, you can use the `KinectSensor.Stop()` method, and to disable the depth stream, use the `KinectSensor.DepthStream.Disable()` method.

We can register to the `KinectSensor.DepthFrameReady` event to process the depth stream data available. The following code snippet defines the details of the `sensor_DepthFrameReady` event handler:

```
void sensor_DepthFrameReady(object sender,
DepthImageFrameReadyEventArgs e)
{    using (DepthImageFrame depthFrame = e.OpenDepthImageFrame())
     {    if
        (depthFrame != null)
         {
           // Copy the pixel data from the image to the pixels array
           depthFrame.CopyDepthImagePixelDataTo(this.depthPixels);

//convert the depth pixels to colored pixels
ConvertDepthData2RGB(depthFrame.MinDepth, depthFrame.MaxDepth);

this.depthBitmap.WritePixels(
             new Int32Rect(0, 0, this.depthBitmap.PixelWidth, this.
depthBitmap.PixelHeight),
             this.colorDepthPixels,
             this.depthBitmap.PixelWidth * sizeof(int),0);}}}
```

The data we obtain for each single `DepthFrame` is copied to the `DepthImagePixel[] depthPixels` array, which holds the depth value of every single pixel in the `depthFrame`. We convert the depth pixels array to the `byte[] colorDepthPixels` using our custom `void ConvertDepthData2RGB(int minDepth, int maxDepth)` method. We finally display the `colorDepthPixels` using the `depthBitmap.WritePixels` method.

 The same consideration about performances and the same polling technique we develop for the color stream data manipulation can be applied to the depth stream data too. As an alternative to subscribing to the `DepthFrameReady` event, we can obtain the current depth frame using the public `DepthImageFrame OpenNextFrame(int millisecondsWait);` method of the `DepthImageStream` class.

DepthRange – the default and near mode

Using the `Int32 DepthRange Range` property of the `DepthImageStream` class, we can select the two available depth range modes:

- `DepthRange.Near`: The Kinect sensor captures with high level of reliability depth points within a range of 0.4 to 3 m

- `DepthRange.Default`: The Kinect sensor captures with high level of reliability depth points within a range of 0.8 to 4 m

The `DepthImageStream.MaxDepth` and `DepthImageStream.MinDepth` properties provide the minimum and maximum reliable depth value according to the `DepthRange` we select.

Extended range

The Kinect sensor is able to capture depth points even outside of the depth ranges defined previously. In this case, the level of reliability of the depth data is decreased.

In the following code, we convert the depth pixels to colored pixels, highlighting:

- Yellow: The point where the depth information is not provided

- Red: All the points that are closer than the `Min` depth dictated by the current `DepthRange`

- Green: All the points that are more than the `Max` depth dictated by the current `DepthRange`

- Blue: Ranges all the others

Using the `this.depthPixels.Max(p => p.Depth)` statement, we can notice that the Kinect sensor is able to render points well over the `MaxDepth`. In an appropriate environment this value can easily reach 10 meters.

```
void ConvertDepthData2RGB(int minDepth, int maxDepth)
{   int colorPixelIndex = 0;
    for (int i = 0; i < this.depthPixels.Length; ++i)
    {   // Get the depth for this pixel
        short depth = depthPixels[i].Depth;
        if (depth == 0) // yellow points
        {   this.colorDepthPixels[colorPixelIndex++] = 0;
            this.colorDepthPixels[colorPixelIndex++] = 255;
            this.colorDepthPixels[colorPixelIndex++] = 255;}
        else
```

```
{     // Write out blue byte
      this.colorDepthPixels[colorPixelIndex++] = (byte)depth;
      // Write out green byte - full green for > maxdepth
      this.colorDepthPixels[colorPixelIndex++] = (byte)(depth >=
      maxDepth ? 255 : depth >> 8);
      // Write out red byte - full red for < mindepth
      this.colorDepthPixels[colorPixelIndex++] = (byte)(depth <=
      minDepth ? 255 : depth >> 10); }
  // If we were outputting BGRA, we would write alpha
     here.

      ++colorPixelIndex; }
   //establish the effective maxdepth for each single frame
   this.dataContext.MaxDepth = this.depthPixels.Max(p => p.Depth); }
```

 The complete code for the Depth stream data manipulation is provided in the CODE_02/DepthStream Visual Studio solution.

Mapping from the color frame to the depth frame

In order to map depth coordinate spaces to color coordinate spaces and vice versa, we can utilize three distinct APIs. The CoordinateMapper.MapDepthFrameToColorFrame and CoordinateMapper.MapColorFrameToDepthFrame enable us to map the entire image frame. The CoordinateMapper.MapDepthPointToColorPoint API is used for mapping one single point from the depth space to the color space. We suggest referring the MSDN for a detailed explanation of the APIs.

In this paragraph, we will list and document the essential steps for mapping a depth stream to the color stream. The CODE_02/CoordinateMapper example attached to this chapter will provide all the additional details.

KinectSensor.AllFramesReady is the event that the sensor fires when all the new frames for each of the sensor's active streams are ready.

We can register to this event to process the streams data available and implement the related event handler using the this.sensor.AllFramesReady += this. sensor_AllFramesReady statement. We initialize the depth and color data and maps using the sensor.CoordinateMapper.MapDepthFrameToColorFrame (DepthImageFormat depthImageFormat, DepthImagePixel[] depthPixels, ColorImageFormat colorImageFormat, ColorImagePoint[] colorPoints) API.

The following code snippet defines the details of the sensor_AllFramesReady event handler:

```
void sensor_AllFramesReady(object sender, AllFramesReadyEventArgs e)
{
   rgbReady = false; depthReady = false;

   using (ColorImageFrame colorFrame = e.OpenColorImageFrame()){
     if (colorFrame != null) {

        //Copy the pre-pixel color data to a pre-allocated array
        colorFrame.CopyPixelDataTo(colorPixels);
        rgbReady = true;}}

    using (DepthImageFrame depthFrame = e.OpenDepthImageFrame()){
        if (depthFrame != null) {

  //Copy the pre-pixel depth data to a pre-allocated array
  depthFrame.CopyDepthImagePixelDataTo(depthPixels);
  mappedDepthLocations = new ColorImagePoint[depthFrame.
  PixelDataLength];
  depthReady = true;}}
    if (rgbReady && depthReady){

   // Coping color image into bitMapBits
   for (int i = 0; i < colorPixels.Length; i += 4){
       bitMapBits[i + 3] = 255;                    //ALPHA
       bitMapBits[i + 2] = colorPixels[i + 2]; //RED
       bitMapBits[i + 1] = colorPixels[i + 1]; //GREEN
       bitMapBits[i] = colorPixels[i];             //BLUE
}

//Maps the entire depth frame to color space.
   this.sensor.CoordinateMapper.MapDepthFrameToColorFrame(
       this.sensor.DepthStream.Format, depthPixels,
       this.sensor.ColorStream.Format, mappedDepthLocations);
    for (int i = 0; i < depthPixels.Length; i++){
        int distance = depthPixels[i].Depth;

//Overlay if distance > 400mm and <1200mm
if (  (distance > sensor.DepthStream.MinDepth) && (distance < 1200)){
        ColorImagePoint point = mappedDepthLocations[i];
        int baseIndex = (point.Y * 640 + point.X) * 4;

/*the point near the edge of the depth frame  correspond to a
pixel beyond the edge of the color frame. We verify that the point
coordinates lie within the color image. */
if (  (point.X >= 0 && point.X < 640) &&
      (point.Y >= 0 && point.Y < 480)){
```

```
//Red overlay + depth image + grid
   bitMapBits[baseIndex + 2] =
   (byte)((bitMapBits[baseIndex + 2] + 255) >> 1);
}}}
//draw the WritableBitmap
  bitMap.WritePixels(new Int32Rect(0, 0,
    bitMap.PixelWidth, bitMap.PixelHeight),
    bitMapBits, bitMap.PixelWidth * sizeof(int), 0);
    this.mappedImage.Source = bitMap;}}
```

For a green overlay without a grid we could use: `bitMapBits[baseIndex + 1] = (byte)((bitMapBits[baseIndex] + 255) >> 1`. For a simple blue overlay without depth data we could use: `bitMapBits[baseIndex] = 255;`

In the following picture, the man is overlapped by a red color and a grid because he is located between 40 cm and 1.2 m from the sensor. We can notice that there isn't an overlay on a portion of the right hand and forearm because depth and color frames come from different sensors, for this pixels date in the two frames may not always line up exactly.

Overlapping entities located between 40 cm and 1.2 m with a red grid

With the `CordinateMapper` API we could easily implement a background subtraction technique with full motion tracking. If necessary, we can also map depth data on frames captured with external full HD color camera for enhanced green screen movie studio capabilities.

Summary

In this chapter we learned how to start building a Kinect project using Visual Studio, and we focused on how to handle color streams and depth streams.

We managed the `KinectSensors_StatusChanged` event handler to retrieve the potential active Kinect sensor and we learned how to start the sensor using the `KinectSensors.Start()` method.

Thanks to the `KinectSensors.ColorStream.Enable(ColorImageFormat)` and the event handler attached to the `KinectSensors.ColorFrameReady`, we started to manipulate the color stream data provided by the Kinect sensor. We then had some fun applying custom effects to the color stream data.

We went through scenarios where retrieving color frames using the `KinectSensors.ColorFrameReady` could cause issues, and we introduced the *polling* technique using the `ColorImageFrame OpenNextFrame (int millisecondsWait)` API instead.

We introduced the SDK V 1.6 new API `ColorStream.CameraSettings` for tuning the images provided by the Kinect sensor.

Using the `ColorImageFrame.Format` property, we detailed the different color image formats handled by the sensor and the detailed actions we need to implement for handling raw formats such as Bayer and YUV.

We then looked at the depth stream data and managed the same listening to the `DepthFrameReady` event. Handling the `DepthRange Range` property, we then managed the depth of the pixels in the `DepthImageFrame`.

Using the `CoordinateMapper.MapDepthFrameToColorFrame` API we calibrated depth and color frames. We also learned how to map depth data on color image in order to enhance an object at a predefined distance or eventually remove the background.

In the next chapter we will explore the skeleton stream and its features. In particular we will walk through how to manage the skeleton stream in order to associate user's gestures with actions.

Before jumping to the next chapter we encourage you to develop all the applications of this chapter. The techniques and the process applied in this chapter are substantial for the exercises developed in the next chapter.

3
Skeletal Tracking

Skeletal tracking allows applications to recognize people and follow their actions. Skeletal tracking combined with gesture-based programming enables applications to provide a natural interface and increase the usability and ease of the application itself.

In this chapter we will learn how to enable and handle the skeleton data stream. For instance, we will address the following:

- Tracking users by analyzing the skeleton data streamed by Kinect and mapping them to the color stream
- Understanding what joints are and which joints are tracked in the near and seated mode
- Observing the movements of the tracked users to detect simple actions

Mastering the skeleton data stream enables us to implement an application by tracking the user's actions and to recognize the user's gestures.

The Kinect sensor, thanks to the IR camera, can recognize up to six users in its field of view. Of these, only up to two users can be fully tracked, while the others are tracked from one single point only, as demonstrated in the following image:

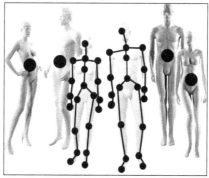

Tracking up to six users in the field of view

Tracking users

The application flow for tracking users is very similar to the process we described in the color frame and depth frame management:

1. Firstly, we need to ensure that at least one Kinect sensor is connected.

2. Secondly, we have to enable the stream (in this case the skeleton one).

3. And finally, we need to handle the frames that the sensor is streaming through the relevant SDK APIs.

In this chapter we will mention only the code that is relevant to skeletal tracking. The source code attached to the book does include all the detailed code and we can refer to the previous chapter to refresh ourselves on how to address step 1.

To enable the skeleton stream, we simply invoke the `KinectSensor.SkeletonStream.Enable()` method.

The Kinect sensor streams out in the skeleton stream's skeleton tracking data. This data is structured in the `Skeleton` class as a collection of joints. A **joint** is a point at which two skeleton bones are joined. This point is defined by the `SkeletonPoint` structure, which defines a 3D position—or point defined in meters by the three values (x,y,z)—in the skeleton space. We have up to twenty joints per single skeleton. A detailed list of the joint types is defined by the `JointType` enumeration at `http://msdn.microsoft.com/en-us/library/microsoft.kinect.jointtype.aspx`.

We are going to store the skeleton data in the `private Skeleton[] skeletonData` array that we size as per the `sensor.SkeletonStream.FrameSkeletonArrayLength` property. This property provides the total length of the skeleton data buffer for the `SkeletonFrame` class and enables skeleton tracking to fully track active skeletons and/or track the location of active skeletons.

We enable our application to listen to and manage the skeleton stream defining the `void sensor_AllFramesReady(object sender, AllFramesReadyEventArgs e)` event handler attached to the `this.sensor.AllFramesReady` event.

The following code snippet summarizes the necessary steps to enable the skeleton stream:

```
//handle the status changed event for the current sensor.
//All the available status value are defined in the Microsoft.Kinect.
KinectStatus enum
void KinectSensors_StatusChanged(object sender, StatusChangedEventArgs
e)
{
```

```
    //select the first (if any available) connected Kinect Sensor from
the KinectSensor.KinectSensors collection
    this.sensor = KinectSensor.KinectSensors.FirstOrDefault(s =>
s.Status == KinectStatus.Connected);

    if (null != this.sensor)
    {//enable the skeleton stream
        sensor.SkeletonStream.Enable();

        // Allocate Skeleton data
        skeletonData = new Skeleton[sensor.SkeletonStream.
FrameSkeletonArrayLength];

        // subscribe to the event raised when all frames are ready
        this.sensor.AllFramesReady += sensor_AllFramesReady;

        // Start the sensor
        try
        {
            this.sensor.Start();}
        catch (IOException)
        {
            this.sensor = null; }
    } }
```

As we have noticed, we subscribed to the AllFramesReady event, which is raised when all the frames (color, depth, and skeleton) are ready. We could rather subscribe to the SkeletonFrameReady event, which is raised when only the skeleton frame is ready. As we will see soon, we opted for the AllFrameReady event because in our example, we need to handle both the skeleton and the color frames.

In this example we will manage the skeleton stream reacting to the frame ready event. We could apply the same consideration debated for the color frame and approach skeleton tracking using the polling technique. To do so, we should leverage the SkeletonStream. OpenNextFrame() method instead of subscribing to the AllFramesReady event or to the SkeletonFrameReady event.

At this stage the code written in the sensor_AllFramesReady event handler should:

- Handle the color stream data
- Handle the skeleton stream data
- Visualize the skeleton drawing color overlapping the color frame

The following code snippet embeds all the activities aforementioned:

```
/// <summary>
/// manage the entire stream data received from the sensor
/// </summary>
/// <param name="sender"></param>
/// <param name="e"></param>
void sensor_AllFramesReady(object sender, AllFramesReadyEventArgs e)
{
    using (ColorImageFrame colorFrame = e.OpenColorImageFrame())
    {   if (colorFrame != null)
        {   //copy the color frame's pixel data to the array
            colorFrame.CopyPixelDataTo(this.colorPixels);

            //draw the WritableBitmap
            this.colorBitmap.WritePixels(
                    new Int32Rect(0, 0, this.colorBitmap.PixelWidth,
this.colorBitmap.PixelHeight),
                    this.colorPixels, this.colorBitmap.PixelWidth *
colorFrame.BytesPerPixel, 0);
        }   }

    //handle the Skeleton stream data
    using (SkeletonFrame skeletonFrame = e.OpenSkeletonFrame())
    // Open the Skeleton frame
    {   if (skeletonFrame != null && this.skeletonData != null)
        // check that a frame is available
        {
            skeletonFrame.CopySkeletonDataTo(this.skeletonData);
            // get the skeletal information in this frame
        }   }

    //draw the output
    using (DrawingContext dc = this.drawingGroup.Open())
    {
        // draw the color stream output
        dc.DrawImage(this.colorBitmap, new Rect(0.0, 0.0, RenderWidth,
RenderHeight));

        //draw the skeleton stream data
        DrawSkeletons(dc);

        // define the limited area for rendering the visual outcome
        this.drawingGroup.ClipGeometry = new RectangleGeometry(new
Rect(0.0, 0.0, RenderWidth, RenderHeight));
    }}
```

For all the explanations related to the color stream data and frame, we can refer to the previous chapter. Let's now focus on the skeleton data stream and how we visualize them overlapping the color frame.

Copying the skeleton data

Thanks to the `SkeletonFrame.CopySkeletonDataTo` method, we can copy the skeleton data to our `skeletonData` array, where we store each skeleton as collection of the joints.

We can draw the skeleton data overlapping the color frame on the screen thanks to an instance of the `System.Windows.Media.DrawingContext` class. We obtain this instance calling the `Open()` method of the `System.Windows.Media.DrawingGroup` class.

There are certainly other ways we could obtain the graphical result. Having said that, the `DrawingGroup` class provides a handy solution to our problem where we need to handle a collection of bones and joints that can be activated upon as a single image.

`RenderWidth` and `RenderHeight` are two double constants set to `640.0f` and `480.0f`. We use them to handle the width and height dimensions of the image we display.

The following code snippet initializes the `DrawingImage imageSource` and `DrawingGroup drawingGroup` variables we use for displaying the graphical outcome of this chapter's example:

```
this.drawingGroup = new DrawingGroup();

// Create an image source that we can use in our image control
this.imageSource = new DrawingImage(this.drawingGroup);

// Display the drawing using our image control
imgMain.Source = this.imageSource;
```

For drawing the skeletons, we loop through the entire skeleton data and we render it skeleton by skeleton. For the skeletons that get fully tracked, we draw a complete skeleton composed by bones and joints. For the skeletons that are not able to be fully tracked, we draw a single ellipse only to highlight their position. We highlight when a user moves to the edge of the field of view. This provides a visual feedback indicating the user skeleton has been clipped:

```
/// <summary>
/// Draw the skeletons defined in the skeleton data
/// </summary>
```

```
/// <param name="drawingContext">dc used to design lines and ellipses
representing bones and joints</param>
private void DrawSkeletons(DrawingContext drawingContext)
{
    foreach (Skeleton skeleton in this.skeletonData)
    {   if (skeleton != null)
        {
            // Fully Tracked skeleton
            if (skeleton.TrackingState == SkeletonTrackingState.
Tracked)
            {
                DrawTrackedSkeletonJoints(skeleton.Joints,
drawingContext);   }
            // Recognized position of the skeleton
            else if (skeleton.TrackingState == SkeletonTrackingState.
PositionOnly)
            {
                DrawSkeletonPosition(skeleton.Position,
drawingContext);   }

            //handle clipped edges
            RenderClippedEdges(skeleton, drawingContext);
    }   }  }
```

We render the fully tracked skeletons using lines to represent bones and ellipses to represent joints. A section of the body is defined as a set of bones and their related joints. The following code snippet highlights the mechanism used to render the head and shoulders. We could apply the same mechanism to render the left arm, the right arm, the body, the left leg, and the right leg:

```
/// <summary>
/// Draw the skeleton joints successfully fully tracked
/// </summary>
/// <param name="jointCollection">joint collection to draw</param>
/// <param name="drawingContext">design the graphical output</param>
private void DrawTrackedSkeletonJoints(JointCollection
jointCollection, DrawingContext drawingContext)
{
    // Render Head and Shoulders
    DrawBone(jointCollection[JointType.Head],
jointCollection[JointType.ShoulderCenter], drawingContext);
    DrawBone(jointCollection[JointType.ShoulderCenter],
jointCollection[JointType.ShoulderLeft], drawingContext);
    DrawBone(jointCollection[JointType.ShoulderCenter],
jointCollection[JointType.ShoulderRight], drawingContext);
```

```
    // Render other bones...

    //Render all the joints
    foreach (Joint singleJoint in jointCollection)
    {
        DrawJoin(singleJoint, drawingContext);
    } }
```

We render a skeleton identified with its position only using a single azure-colored ellipse, as defined in the following code snippet:

```
/// <summary>
/// Draw the skeleton position only
/// </summary>
/// <param name="skeletonPoint">skeleton single point</param>
/// <param name="drawingContext">dc used to design the graphical
output</param>
private void DrawSkeletonPosition(SkeletonPoint skeletonPoint,
DrawingContext drawingContext)
{
    drawingContext.DrawEllipse(Brushes.Azure, null, this.SkeletonPoint
ToScreen(skeletonPoint), 2, 2); }
```

The following code demonstrates how we can provide a visual feedback when the user moves to the edge of the field of view. Thanks to the `Skeleton.ClippedEdges.HasFlag` method, the skeletal tracking system provides a feedback whenever the user skeleton has been clipped on a given edge:

```
/// <summary>

/// Highlights the edge where the skeleton data have been clipped

/// </summary>

/// <param name="skeleton">single skeleton</param>

/// <param name="drawingContext">dc used to design the graphical
output</param>

private void RenderClippedEdges(Skeleton skeleton, DrawingContext
drawingContext)

{   //tests wherever the user skeleton has been clipped or not
    if (skeleton.ClippedEdges.HasFlag(FrameEdges.Bottom))

    {   // colors the bottom border when the user is reaching it
        drawingContext.DrawRectangle(
```

```
                    Brushes.Red,

                    null,

                    new Rect(0, RenderHeight - 10, RenderWidth, 10));

    }

        //manage the other edges
    }
```

As stated previously, we intend a bone to be a line connecting two adjacent joints. The single joint can assume a `TrackingState` value defined by the `JointTrackingState` enum: `NotTracked`, `Inferred`, and `Tracked`. We define a bone as *tracked* if and only if both the joints have `TrackingState` equal to `JointTrackingState.Tracked`. We define a bone as *non-tracked* if at least one of its joints has `TrackingState` equal to `JointTrackingState.Inferred`. We are not able to render the bone if any of its joints has `TrackingState` equal to `JointTrackingState.NotTracked`:

```
/// <summary>
/// draw a bone as line between two given joints
/// </summary>
/// <param name="jointFrom">starting joint of the bone</param>
/// <param name="jointTo">ending joint of the bone</param>
/// <param name="drawingContext">dc used to design the graphical
output</param>
private void DrawBone(Joint jointFrom, Joint jointTo, DrawingContext
drawingContext)
{   if (jointFrom.TrackingState == JointTrackingState.NotTracked ||
    jointTo.TrackingState == JointTrackingState.NotTracked)
    {
        return; // nothing to draw, one of the joints is not tracked
    }

    if (jointFrom.TrackingState == JointTrackingState.Inferred ||
    jointTo.TrackingState == JointTrackingState.Inferred)
    {
        // Draw thin lines if either one of the joints is inferred
        DrawNonTrackedBoneLine(jointFrom.Position, jointTo.Position,
drawingContext);
    }

    if (jointFrom.TrackingState == JointTrackingState.Tracked &&
    jointTo.TrackingState == JointTrackingState.Tracked)
    {
        // Draw bold lines if the joints are both tracked
        DrawTrackedBoneLine(jointFrom.Position, jointTo.Position,
drawingContext);
    }}
```

We draw the bone simply by calling the `DrawingContext.DrawLine` method. We can use two different colors for differentiating between tracked bones and non-tracked bones. For example, we can define `Pen trackedBonePen = new Pen(Brushes.Gold, 6)` for tracked bones. The following method defines the way we render tracked bones:

```
/// <summary>
/// draw a line representing a tracked bone
/// </summary>
/// <param name="skeletonPointFrom">starting point of the bone</param>
/// <param name="skeletonPointTo">ending point of the bone</param>
/// <param name="drawingContext">dc used to design the graphical
output</param>
private void DrawTrackedBoneLine(SkeletonPoint skeletonPointFrom,
SkeletonPoint skeletonPointTo, DrawingContext drawingContext)
{
    drawingContext.DrawLine(this.trackedBonePen, this.SkeletonPointT
oScreen(skeletonPointFrom), this.SkeletonPointToScreen(skeletonPoint
To));
}
```

Similarly, we can draw the joints as ellipses and differentiate those with `TrackingState` equal to `JointTrackingState.Tracked` from those with `TrackingState` equal to `JointTrackingState.Inferred` or `JointTrackingState.NotTracked`. The following code snippet indicates how we can render the joint and adjust it according to the joints' `TrackingState`:

```
    if (singleJoint.TrackingState == JointTrackingState.NotTracked)
    {
        return; // nothing to draw
    }
// singleJoint is the joint to draw
    if (singleJoint.TrackingState == JointTrackingState.Inferred)
    {
        DrawNonTrackedJoint(singleJoint, drawingContext);
        // Draw thin ellipse if the joint is inferred
    }

// drawingContext is the dc used to design the graphical

    if (singleJoint.TrackingState == JointTrackingState.Tracked)
    {
        DrawTrackedJoint(singleJoint, drawingContext);
        // Draw bold ellipse if the joint is tracked
    }

private void DrawTrackedJoint(Joint singleJoint, DrawingContext
drawingContext)
```

```
{
    drawingContext.DrawEllipse(
                this.trackedJointBrush,
                null,
                this.SkeletonPointToScreen(singleJoint.Position),
                10, 10);
}
```

To visualize the single skeletons overlapping the color image in the right position, we utilize the `CoordinateMapper.MapSkeletonPointToColorPoint` method, which maps a point from skeleton space to color space:

```
/// <summary>
/// Maps a SkeletonPoint to lie within our render space and converts
to Point
/// </summary>
/// <param name="skelpoint">point to map</param>
/// <returns>mapped point</returns>
private Point SkeletonPointToScreen(SkeletonPoint skelpoint)
{
    // Convert point to color space.
    // We are assuming our output resolution to be 640x480.
    ColorImagePoint colorPoint = this.sensor.CoordinateMapper.
MapSkeletonPointToColorPoint(skelpoint, ColorImageFormat.
RgbResolution640x480Fps30);
    return new Point(colorPoint.X, colorPoint.Y);
}
```

We are now ready: our skeletons overlap the color data stream and we can take a funny x-ray of ourselves. The full list of joints is detailed in the `JointType` enumeration available online at `http://msdn.microsoft.com/en-us/library/microsoft.kinect.jointtype.aspx`. The joint state is detailed in the `JointTrackingState` enumeration available at `http://msdn.microsoft.com/en-us/library/microsoft.kinect.jointtrackingstate.aspx`.

The Kinect sensor in its skeletal tracking mode by default selects the first two recognized users in the field of view. We can use the `AppChoosesSkeletons` and `ChooseSkeletons` members of the `SkeletonStream` class to actively choose in the application which skeleton to track among the six users recognized in the field of view.

We may decide to track the closest skeleton or the skeleton that falls in a predefined distance interval. The source code attached to this chapter defines a simple routine for tracking the closest skeleton.

The remaining four skeletons are tracked highlighting the `HipCenter` (center, between hips) joint only.

Default and Seated mode

As we saw in the previous chapter, the Kinect for Windows SDK provides a near-range feature in order to track people close to the sensor.

First of all, in order to activate the near tracking mode we need to enable the near-range feature by setting the `sensor.DepthStream.Range` property to `DepthRange.Near`; then by setting the `sensor.SkeletonStream` property to `true`.

This mode usually, in addition to tracking users in the range 0.4 – 0.8 m, allows for greater accuracy up to 3 m than the Default mode.

For scenarios where the user to be tracked is seated, or the lower part of his/her body is not entirely visible to the sensor, we can enable the Seated mode by setting the `sensor.SkeletonStream.TrackingMode` property to `SkeletonTrackingMode.Seated`. With this mode, the APIs track only the upper-body part's joints and will get a `NotTracked` status for all of the remaining joints.

The following image highlights the twenty joint points for the Default mode and joints ten joint points for the Seated mode:

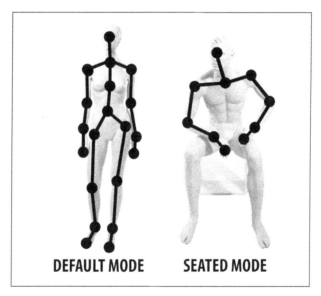

DEFAULT MODE   SEATED MODE

Tracking details for a skeleton in Default and Seated mode

 It should have been noticeable that enabling the Seated mode decreases the skeleton frame throughput and hence decreases the performances. We recommend not to use the Seated mode to filter out the lower-body joint points.

Detecting simple actions

Let's see now how we can enhance our application and leverage the Kinect sensor's **Natural User Interface** (**NUI**) capabilities.

We implement a manager that, using the skeleton data, is able to interpret a body motion or a posture and translate the same to an action as "click". Similarly, we could create other actions as "zoom in". Unfortunately, the Kinect for Windows SDK does not provide APIs for recognizing gestures, so we need to develop our custom gesture recognition engine.

Gesture detection can be relatively simple or intensely complex depending on the gesture and the environment (image noise, scene with more users, and so on).

In literature there are many approaches for implementing gesture recognition, the most common ones are as follows:

- **A neural network** that utilizes the weighted networks (*Gestures and neural networks in human-computer interaction, Beale R* and *Alistair D N E*)
- **A DTW** that utilizes the Dynamic Time Warping algorithm initially developed for the speech recognition and signal processing (*Space-Time Gestures, T. Darrell* and *A. Prentland; Spatial-Temporal Features by Image Registration and Warping for Dynamic Gesture Recognition, Y. Huang, Y ZHU, G. XU, H. Zhang*)
- **The Adaptive Template method** (*Adaptive Template Method for Early Recognition of Gestures, K. Kawashima, A. Shimada, H. Hagahara,* and *R. Taniguchi*)
- **HMMs** that utilize statistical classification (*Hidden Markov Model for Gesture Recognition, J. Yang* and *Y. Xu*)
- **The Hybrid approach** that utilizes a combination of the previously mentioned approaches

In this chapter we are going to develop the manager for gesture recognition based on an algorithmic approach that considers a gesture as a sequence of postures defined by the position of the tracked joints. This algorithm is based on a simplification of the Adaptive Template method that uses the skeletal tracking data provided by the SDK. Understanding the complexity of all the approaches previously listed, we decided to use the algorithm approach because it is simple and suits the scope of this book.

We are going to develop an example where we select an area of the scene captured by the Kinect color camera of raising our left hand, and then we can drag the selected area of moving the right hand. The scope of this example is to demonstrate how we can translate a simple gesture in to a command for the application. Hence, this example will demonstrate how we can use the Kinect sensor for leveraging NUI in our applications.

The algorithmic approach enables us to define and easily distinguish simple gestures. It is optimal for tracking uniform movements. The algorithmic approach does face a few challenges on recognizing more complex gestures such as the *swing* movement or drawing a circle.

Every single gesture can be decomposed in to two or more sections (or segments). A **section** is a posture at a given time. A single **posture** is defined as the location of a given joint in respect to other joints. For example, the gesture of raising the left hand could be defined by the two sections: we start by having our left and right hands alongside the body and lower than the shoulders; we finish with the left hand higher than the left shoulder while the right hand is still lower than the right shoulder.

For simplicity we will decompose our initial `SelectionHandLeft` gesture in to two sections that we are defining as the initial section and the final section. To recognize the `SelectionHandLeft` gesture we have to first recognize the initial section and then the final one. Every single section is validated. The single gesture is validated if and only if all the sections composing the gesture itself are validated. Once the gesture has been validated, the manager notifies all the observers with the gesture recognition event.

The code provided in the following code snippet is a custom code for a single section defined as a class that implements the `IgestureSection` interface:

```
//GESTURE'S SECTION INTERFACE
interface IgestureSection
{
    GestureSectionCheck Check(Skeleton skeleton);
}
```

The initial and final sections of the "Selection Hand Left:" gesture are defined by the custom code classes, `SelectionGestureHandLeftSTARTsection` and `SelectionGestureHandLeftENDsection`, described in the following code snippets:

```
class SelectionGestureHandLeftSTARTsection : IgestureSection
{
/// <summary> Validate a gesture's section
/// </summary>
/// <param name="skeleton">skeleton stream data</param>
/// <returns>'OK' if gesture is validate, 'KO' otherwise</returns>
public GestureSectionCheck Check(Skeleton skeleton)
  { if (skeleton.Joints[JointType.HandLeft].Position.X <
skeleton.Joints[JointType.ShoulderLeft].Position.X && skeleton.
Joints[JointType.HandRight].Position.Y <
    skeleton.Joints[JointType.ShoulderRight].Position.Y &&
    skeleton.Joints[JointType.HandLeft].Position.Z >
```

```
    skeleton.Joints[JointType.ShoulderLeft].Position.Z-0.30f)
        {
            return GestureSectionCheck.ok;
        }
        return GestureSectionCheck.ko;
    }
}

class SelectionGestureHandLeftENDsection : IgestureSection
{public GestureSectionCheck Check(Skeleton skeleton)
    {if (skeleton.Joints[JointType.HandLeft].Position.X <
    skeleton.Joints[JointType.ShoulderLeft].Position.X &&
    skeleton.Joints[JointType.HandRight].Position.Y <
    skeleton.Joints[JointType.ShoulderRight].Position.Y &&
    skeleton.Joints[JointType.HandLeft].Position.Z <
    skeleton.Joints[JointType.ShoulderLeft].Position.Z-0.30f)
        {
            return GestureSectionCheck.ok;
        }
        return GestureSectionCheck.ko;
    }
}
```

> Splitting a single gesture into discrete sections increases the reliability of selecting and recognizing the right gesture.
>
> We can improve the gesture recognition algorithm by increasing the level of the status managed by the Check method to allow a more detailed analysis of the intermediate movements.

The following code snippet defines our base class for the gestures:

```
class Gesture
{
    private IgestureSection[] gestureSections;
    private int counterGestureSection = 0;
    private int counterFrame = 0;
    private GestureType gestureType;
    public event EventHandler<GestureEventArgs> GestureRecognized;
    public Gesture(IgestureSection[] gestureSections,  GestureType
gestureType)
    {
        this.gestureSections = gestureSections;
        this.gestureType = gestureType;
    }
```

```
    public void Update(Skeleton Data)
    {GestureSectionCheck check =
this.gestureSections[this.counterGestureSection].Check(Data);
        if (check == GestureSectionCheck.ok)
        {
            if (this.counterGestureSection + 1 < this.gestureSections.
Length)
            {
                this.counterGestureSection++;
                this.counterFrame = 0;
            }
            else
            {
                if (this.GestureRecognized != null)
                {
                    this.GestureRecognized(this,
                new GestureEventArgs(this.gestureType));
                    this.Reset();
                }
            }
        }
        else if (check == GestureSectionCheck.ko || this.counterFrame
== 60)
        {
            this.Reset();

        }
        else
        {
            this.counterFrame++;
        }
    }

    public void Reset()
    {
        this.counterGestureSection = 0;
        this.counterFrame = 0;
    }

}

    public enum GestureType
    {
        NoGesture,
        SelectionGestureHandLeft
    //ADD OTHER GESTURE TYPE
    }

    //STATE OF GESTURE'S SECTION CHECK
    public enum GestureSectionCheck
```

```
{
    ko,
    ok
}


class GestureEventArgs:EventArgs
{
    public GestureType GestureType
    {
        get;
        set;
    }
    public GestureEventArgs(GestureType gestureType)
    {
        this.GestureType = gestureType;
    }
}
```

In the following code we define the gesture manager. In the GestureManager class construct we define, in a single collection, all the different gestures that the manager is going to handle and all the sections composing a single gesture. As previously stated, in our example, the "Selection Hand Left" gesture is composed of two sections only:

```
class GestureManager
{

    private List<Gesture> gestures = new List<Gesture>();

    public event EventHandler<GestureEventArgs> GestureRecognized;

    public GestureManager()
    {
    IgestureSection[] SelectionSectionsHandLeft = new
IgestureSection[2];
    SelectionSectionsHandLeft[0] = new
SelectionGestureHandLeftSTARTsection();
    SelectionSectionsHandLeft[1] = new
SelectionGestureHandLeftENDsection();
    Add(SelectionSectionsHandLeft, GestureType.
SelectionGestureHandLeft);

//ADD HERE OTHER GESTURE
    }
 public void Add(IgestureSection[] gestureSections,GestureType
gestureType)
```

```
{
    Gesture gesture = new Gesture(gestureSections, gestureType);
    gesture.GestureRecognized += gesture_GestureRecognized;
    this.gestures.Add(gesture);
}


void gesture_GestureRecognized(object sender, GestureEventArgs e)
{
    if (this.GestureRecognized != null)
    {
        this.GestureRecognized(this, e);
    }

    ResetAllGestures();
}


public void UpdateAllGestures(Skeleton data)
{
    foreach (Gesture gesture in this.gestures)
    {
        gesture.Update(data);
    }
}


public void ResetAllGestures()
{
    foreach (Gesture gesture in this.gestures)
    {
        gesture.Reset();
    }
}
```

To utilize the gesture manager in our application, we need to declare and instantiate the `private GestureManager gestureManager` variable and define the `gestureManager_GestureRecognized` event handler for the `GestureManager`. `GestureRecognized` event:

```
void gestureManager_GestureRecognized(object sender, GestureEventArgs
e)
{
    switch (e.GestureType)
    {
        case    GestureType.SelectionGestureHandLeft:

        // THING TO DO WHEN THIS GESTURE WAS RECOGNIZED
    // ADD HERE ALL THE GESTURE TO BE MANAGED
        default:
            break;
    }
}
```

Finally, anytime the skeleton data stream is providing a new frame, we need to update the gesture manager and analyze the frame itself to detect potential new gestures or sections. We add the following code snippet that is needed at the end of the `sensor_AllFramesReady` event handler:

```
// update the gesture manager
if (skeletonData != null)
{
foreach (var skeleton in this.skeletonData)
{if (skeleton.TrackingState != SkeletonTrackingState.Tracked)
        continue;
    gestureManager.UpdateAllGestures(skeleton);
}
}
```

Joint rotations

We may face some scenarios where a given action needs to be designed according to **joint rotations**. The Kinect sensor is able to capture the **hierarchical** (rotation of the joint axis with regard to its parent joint) and **absolute** (rotation of the joint axis with regard to the Kinect sensor) joint rotations.

Discussing the joint rotations goes beyond the scope of this book and we recommend you review the SDK v1.6 documentation for a complete description of joint rotations (`http://msdn.microsoft.com/en-us/library/hh973073.aspx`).

During our tests we have been noticing that there is some noise in the joint positions streamed by the skeletal tracking system. An important step for improving the quality of skeletal tracking is to use a noise reduction filter. Applying the filter before the analysis of the skeletal tracking data helps to remove a part of the noise from the joint data. Such filters are called **smoothing filters** and the process is called **skeletal joint smoothing**. A full and in-depth study of skeletal joint smoothing is available in the Microsoft White Paper at `http://msdn.microsoft.com/en-us/library/jj131429.aspx`.

We can certainly share at this stage the smoothing parameters we have been testing for optimizing the joint rotation recognition in some of the proof of concepts developed in the past:

```
// Typical smoothing parameters for the bone orientations:
var boneOrientationSmoothparameters = new TransformSmoothParameters
{    Smoothing = 0.5f,
     Correction = 0.8f,
     Prediction = 0.75f,
     JitterRadius = 0.1f,
     MaxDeviationRadius = 0.1f   };
// Enable skeletal tracking
sensor.SkeletonStream.Enable(boneOrientationSmoothparameters);
```

Using the Kinect sensor as a Natural User Interface device

The source code attached to this chapter provides a fully-functional example, where we demonstrate how simple user's actions can be combined to address a real scenario and utilize the Kinect sensor as a Natural User Interface for a complex application.

In the proposed example, we select a portion of the color camera stream data of raising our left or right hand. The selected portion of the color stream data can then be dragged within the field of view using the other hand (the right hand if we selected using the left one and vice versa).

Summary

In this chapter we learned how to track the skeletal data provided by the Kinect sensor and how to interpret them for designing relevant user actions.

With the example developed in this chapter, we definitely went to the core of designing and developing Natural User Interfaces.

Thanks to the `KinectSensors.SkeletonStream.Enable()` method and the event handler attached to `KinectSensors.AllFramesReady`, we have started to manipulate the skeleton stream data and the color stream data provided by the Kinect sensor and overlap them.

We addressed the `SkeletonStream.TrackingMode` property for tracking users in Default (stand-up) and Seated mode. Leveraging the Seated mode together with the ability to track user actions is very useful for application-oriented people with disabilities.

We went through the algorithmic approach for tracking user's actions and recognizing user's gestures and we developed our custom gesture manager. Gestures have been defined as a collection of movement sections for increasing the reliability of the gesture engine. The gestures dealt with in this chapter are simple but the framework we developed can handle more articulated gestures based on discrete movements. Alternative approaches such as the neural network approach or the template-based approach should be considered in cases where the gestures to track are more complex and cannot be decomposed easily in discrete, well-defined movements. This chapter listed a set of references we could use to understand and explore these alternative approaches.

In the code built on this chapter, together with the full version attached to the book, we demonstrated how we could control the skeleton and color stream data and interact with the objects in the Kinect sensor's field of view. This represents a starting point for delivering an augmented reality experience. We encourage you to enhance the example developed in this chapter. You may want to embed content search capabilities in the application and submit queries related to the objects you interact with.

In the next chapter we will explore the voice tracking data to enhance the example developed in this chapter, and we will develop what is a real **multimodal interface** (voice plus gestures to interact with the application).

Before jumping in to the next reading, we encourage you to develop all the applications in this chapter. You may want to consider the application proposed in this chapter as the starting point to develop an application that can help you to virtually redesign the layout of your room or garage.

4
Speech Recognition

In the previous chapter we saw how to use the Kinect sensor skeleton tracker for providing inputs to our application. In this chapter we will explicate how to use the Kinect sensor's speech recognition capability as an additional natural interface modality in our applications. Speech recognition is a powerful interface that increases the adoption of software solutions by users with disabilities. Speech recognition can be used in working environments where the user can perform his/her job or task away from a traditional workstation.

In this chapter we will cover the following topics:

- The Kinect sensor audio stream data
- Grammars defined by XML files and programmatically
- How to manage the Kinect sensor beam and its angle

The Microsoft Kinect SDK setup process includes the installation of the speech recognition components.

The Kinect sensor is equipped with one array of four microphone devices.

The array of microphones can be handled using the code libraries released by Microsoft since Windows Vista. These libraries include **Voice Capture DirectX Media Object (DMO)** and the **Speech Recognition API (SAPI)**.

In managed code, Kinect SDK v1.6 provides a wrapper extending the Voice Capture DMO. Thanks to the Voice Capture DMO, Kinect provides capabilities such as:

- **Acoustic echo cancellation (AEC)**
- **Automatic gain control (AGC)**
- **Noise suppression**

The Speech Recognition API is the development library that allows us to use the built-in speech recognition capabilities of the operating system while developing our custom application. These APIs can be used with or without the Kinect sensor and its SDK.

Speech recognition

Let's understand what a speech recognition process is and how it works.

The goal of the speech recognition process is to convert vocal commands spoken by the user into actions performed by the application. The speech recognition process is executed by a speech recognizer engine that analyzes the speech input against predefined grammar.

The scope of the speech recognizer engine is to verify that the received speech input is a valid command. A valid command is one that satisfies the syntactic and semantic rules defined by grammar. A valid command recognized by the speech recognizer engine is then converted into actions that the application can execute.

Grammars

Grammar defines all the rules and the logical speech statements we want to apply in our specific situations. In order to accept and process more natural speaking styles and improve the user experience, we should aim to define flexible grammars.

The grammar we use is based on the standard defined by the W3C **Speech Recognition Grammar Specification** Version 1.0 (**SRGS**). This grammar is defined using XML files. The detailed specifications of the grammar and the XML schema are published at http://www.w3.org/TR/speech-grammar.

A simple grammar sample

Let's introduce a simple grammar structured as a list of rules that declares the words and/or phrases used by the speech recognition engine to analyze the speech input.

Our goal is to define a set of commands that the user can enunciate. We list all the commands as words and/or phrases in the grammar file. The speech engine is going to analyze the Kinect sensor audio stream data and map them against the commands list. Once a given command is recognized with sufficient confidence, we then execute the action associated to the command.

The following XML file defines the rules of our simple grammar. The grammar is implemented using two distinct semantic categories: **UP** and **DOWN**. Within any single category we can define one or more as synonymous using the `<item>` node (for example, `up`, `move up`, and `tilt up` are all synonymous and the speech engine treats them in the same way):

```
<grammar version="1.0" xml:lang="en-US" root="rootRule" tag-
format="semantics/1.0-literals" xmlns="http://www.w3.org/2001/06/
grammar">
```

```
    <rule id="rootRule">
      <one-of>
        <item>
          <tag>UP</tag>
          <one-of>
            <item> up </item>
            <item> move up </item>
            <item> tilt up </item>
          </one-of>
        </item>
        <item>
          <tag>DOWN</tag>
          <one-of>
            <item> down </item>
            <item> move down </item>
            <item> tilt down </item>
          </one-o>
        </item>
      </one-of>
    </rule>
</grammar>
```

Instead of using an XML file, we could create grammar
programmatically as demonstrated in the following code snippet:

```
var commands = new Choices();
commands.Add(new SemanticResultValue("up",
"UP"));
commands.Add(new SemanticResultValue("move up",
"UP"));
commands.Add(new SemanticResultValue("tilt up",
"UP"));
commands.Add(new SemanticResultValue("down",
"DOWN"));
// etc.
var gb = new GrammarBuilder { Culture =
ri.Culture };
gb.Append(commands);
var g = new Grammar(gb);
```

The grammar can be amended programmatically too. We may
decide to change the grammar while our application is running.

There are scenarios where we could use rules and semantic definitions to organize grammar's content into logical groupings. This approach is very useful for decreasing the amount of information to be processed. For instance, we could imagine a video game where we need to add or remove words and/or phases per each specific stage or level of the video game.

 Please be aware that in a Visual Studio project, it is not sufficient to include the XML file containing the grammar. In order to consume the grammar, we need to include the XML file as the project's resource. Otherwise the grammar will not be recognized.

We can load the grammar invoking the constructor of the `Microsoft.Speech.Recognition.Grammar` class.

It is possible to load (or unload) simultaneously more than one grammar set in a given speech recognition engine. (Using the `SpeechRecognitionEngine.UnloadAllGrammar()` method we can unload all the grammar sets currently associated to our speech recognition engine).

The Microsoft.Speech library

Even though we could find various similarities between the `System.Speech` library and the `Microsoft.Speech` one, we need to use the latter one as it is optimized for the Kinect sensor. Having said that, it is worth noticing that the `Microsoft.Speech` library's recognition engine doesn't support **DictationGrammar** (dictation model), which is instead supported in `System.Speech`.

Once we have defined the grammar, we can load the same in our speech recognizer engine. Then we need to handle the `SpeechRecognized` event (raised when the speech input has been recognized against one of the semantic categories defined in the grammar) and the `SpeechRecognizedRejected` event (raised when the speech input cannot be recognized against any of the semantic categories).

The following code snippet demonstrates how we initialize the speech recognizer engine. The speech recognizer engine needs to be initialized after streaming data from the Kinect sensor has been started.

For simplicity, we do not report here all the code for starting the Kinect sensor, as it is the same discussed in the previous chapters. Having said that, the source code attached to this chapter includes all the full functioning initialization code.

Let's review the standard approach published by Microsoft at `http://msdn.microsoft.com/en-us/library/jj131035.aspx` for managing speech recognition.

The instance of the `SpeechRecognitionEngine` class is obtained thanks to the `RecognizerInfo` class. The `RecognizerInfo` class provides information about a `SpeechRecognizer` or `SpeechRecognitionEngine` instance. In order to retrieve all the information associated to the Kinect sensor recognizer, we create the `RecognizerInfo` class instance calling the `GetKinectRecognizer` method (detailed in the second code snippet):

```
RecognizerInfo ri = GetKinectRecognizer();

if (null != ri){
    this.speechEngine = new SpeechRecognitionEngine(ri.Id);

    using (var memoryStream =
    new MemoryStream(
    Encoding.ASCII.GetBytes(Properties.Resources.SpeechGrammar)))
            {
                var g = new Grammar(memoryStream);
                speechEngine.LoadGrammar(g);
            }

    speechEngine.SpeechRecognized += SpeechRecognized;
    speechEngine.SpeechRecognitionRejected += SpeechRejected;

    speechEngine.SetInputToAudioStream(
        sensor.AudioSource.Start(),
        new SpeechAudioFormatInfo
        (EncodingFormat.Pcm, 16000, 16, 1, 32000, 2, null));
                speechEngine.RecognizeAsync(RecognizeMode.Multiple);
            }
            else
            {
    //TO DO WHEN NO SPEECH RECOGNIZED
            }
        }
```

The `GetKinectRecognizer` method retrieves the speech recognizer engine parameters that better suit the Kinect sensor audio stream data. The optimal speech recognition engine for the Kinect sensor is obtained from the `SpeechRecognitionEngine.InstalledRecognizers()` collection of speech recognizer engines available on our machine:

```
private static RecognizerInfo GetKinectRecognizer()
{
    foreach (RecognizerInfo recognizer in SpeechRecognitionEngine.
InstalledRecognizers())
    {
        string value;
        recognizer.AdditionalInfo.TryGetValue("Kinect", out value);
```

```
        if ("True".Equals(value, StringComparison.OrdinalIgnoreCase)
&& "en-US".Equals(recognizer.Culture.Name, StringComparison.
OrdinalIgnoreCase))
        {
            return recognizer;
        }
    }

    return null;
}
```

The `SpeechRecognitionEngine.SetInputToAudioStream` method contained in the `Microsoft.Speech.Recognition` namespace assigns the Kinect Sensor audio stream using `sensor.AudioSource.Start()` as the input for the current speech recognizer engine instance.

The parameters' values used in our example for setting `SpeechAudioFormatInfo` are as follows:

- The encoding format is **Pulse Code Modulation (PCM)**
- 16000 samples captured per second
- Every single sample is sized on 16 bits
- We listen to 1 channel
- An average bitrate of 32 KB/s
- 2 bytes for block alignment

The speech recognition can be performed both synchronously and asynchronously. The `RecognizeAysnc` method is for async recognition, where you need an application to be responsive while the speech recognition engine is performing its job. The `Recognize` method is for sync operations and can be used when the responsiveness of the application is not a concern.

The RecognizeMode enum value can be set to either Single or Multiple. The value Single ensures the speech recognizer engine performs only a single operation; whereas setting the value to Multiple will have the recognition to continue to recognize and will not terminate after completion of one speech recognition. In our example, the language and culture used is English en-US. The Kinect sensor supports additional acoustic allowing speech recognition in several locales such as en-GB, en-IE, en-AU, en-NZ, en-CA, French fr-FR, fr-CA, Germany de-DE, and Italian it-IT. The complete list and all the related components — packaged individually — are available at http://www.microsoft.com/en-us/download/details.aspx?id=34809.

For utilizing a different language we need to change the en-US value in the GetKinectRecognizer method and the grammar file. Please note that this setting is detached from the globalization setting defined at operating system level.

The SpeechRecognized event handler handles the result of the speech input analysis and its level of confidence. The SpeechRecognizedEventArgs event argument provides the outcome of the speech recognition through the RecognictionResult Result property.

By testing the SpeechRecognizedEventArgs.Result.Semantics.Value value and assessing SpeechRecognizedEventArgs.Result.Confidence, we can perform the appropriate actions and execute the commands specified in the grammar.

We have to notice that the value of the SpeechRecognizedEventArgs.Result.Confidence property provides the level of confidence the speech bestows on the computed result. The case of Confidence = 1 (100 percent) indicates that the engine is completely confident of what was spoken. On the other end, the case of Confidence = 0 (0 percent), the engine completely lacks confidence.

While performing our example we were in an environment where we could assume a low level of accuracy. For this reason we defined a threshold of 30 percent (0.3). In case we obtain a confidence value lower than the threshold, we reject the result.

It is clear that the threshold is critical to define the level of confidence required to accept the speech recognition process' result. Identifying the right threshold value is a mindful decision we need to take, trading-off elements such as the type of application (its criticality), as well as the environment where the Kinect sensor is operating.

 We strongly recommend performing a tuning session before finalizing the threshold value.

```
private void SpeechRecognized(object sender,
SpeechRecognizedEventArgs e)
{
    // Speech utterance confidence below which we treat speech as if
it hadn't been heard
    const double ConfidenceThreshold = 0.3;

    if (e.Result.Confidence >= ConfidenceThreshold)
    {
        switch (e.Result.Semantics.Value.ToString())
        {
            case "UP":
                tbRecognizer.Text = "Recognized 'UP' @ " + e.Result.
Confidence;
                //we increase the kinect elevation angle of 2 degree
                this.sensor.ElevationAngle = (int)(Math.Atan(previousA
ccelerometerData.Z / previousAccelerometerData.Y) * (180 / Math.PI) +
2);
                break;

            case "DOWN":
                tbRecognizer.Text = "Recognized 'DOWN' @ " + e.Result.
Confidence;
                //we decrease the kinect elevation angle of 2 degree
                this.sensor.ElevationAngle = (int)(Math.Atan(previousA
ccelerometerData.Z / previousAccelerometerData.Y) * (180 / Math.PI) -
2);;
                break;

    }    }}
```

The previous code snippet details our `SpeechRecognized` event handler implementation. We perform the commands related to the matched grammar semantic if and only if the confidence provided by the speech recognizer engine is above the 0.3 threshold value.

We will notice that we do not need to test all the grammar entries, but instead focus on the semantic categories only. The entire related synonyms are handled by the speech recognition engine itself as a whole:

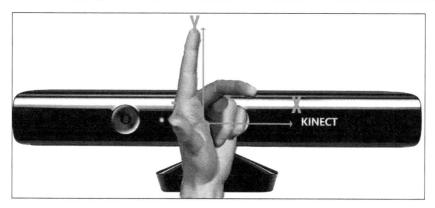

The orientation axis of the Kinect sensor's accelerometer

In our example we attach to the UP and DOWN commands to actions we want to perform against the **Kinect sensor accelerometer**. The Kinect for Windows Sensor is equipped with a 3-axis accelerometer configured for a *2g* range, where *g* is the acceleration due to gravity. The axis' orientation is highlighted in the previous image. The accelerometer enables the sensor to report its current orientation computed with respect to gravity.

Testing the accelerometer data can help us to detect when the sensor is in an unusual orientation. We may be able to use the angle between the sensor and the floor plane and adjust the 3D projection's data in augmented reality scenarios. The accelerometer has a lower limit of 1 degree accuracy. In addition, the accuracy is slightly temperature-sensitive, with up to 3 degrees of drift over the normal operating temperature range. This drift can be positive or negative, but a given sensor will always exhibit the same drift behavior. It is possible to compensate for this drift by comparing the accelerometer's vertical data (the y axis in the accelerometer's coordinate system) and the detected floor plane depth data, if required.

We can control programmatically the Kinect sensor's field of view using the tilt motor in the sensor. The motor can vary the orientation of the Kinect sensor with an angle of +/-27 degrees. The tilt is relative to gravity rather than relative to the sensor base. An elevation angle of zero indicates that the Kinect is pointing perpendicular to gravity.

In our example we combine three different Kinect sensor capabilities to simulate a complex scenario:

- Speech recognition enables us to issue commands to the Kinect sensor using voice as natural interface—this is demonstrated by the semantics UP and DOWN in our SpeechRecognized event handler implementation.

- The Kinect sensor's accelerator provides the current value for the elevation angle of the tilt monitor. The following formula computes the current elevation angle:

```
Math.Atan(previousAccelerometerData.Z /
previousAccelerometerData.Y) * (180 / Math.PI)
```

Where:

  - ○ `previousAccelerometerData` is the latest accelerometer data captured using the `KinectSensor.AccelerometerGetCurrentReading` method

  - ○ In the Kinect sensor's orientation is a right-handed coordinate system centered on the sensor with positive z in the direction to where the sensor is pointing

- We adjust the Kinect sensor's orientation using the `KinectSensor.ElevationAngle` property.

> At the time of writing this book, the `KinectSensor.ElevationAngle` getter had a bug, and it would raise a system exception every time we tried to access it. Hence, to increase (decrease) the Kinect sensor's elevation angle, it is not possible to use the the simple statement `KinectSensor.ElevationAngle += value`.

In addition to the `SpeechRecognized` event handler implementation, we recommend to implement the `SpeechRejected` event handler too, and provide the relevant feedback that the recognition has been rejected:

```
/// <summary>
/// Handler for rejected speech events.
/// </summary>
/// <param name="sender">object sending the event.</param>
/// <param name="e">event arguments.</param>
private void SpeechRejected(object sender,
SpeechRecognitionRejectedEventArgs e)
{
    //Provide a feedback of recognition rejected
}
```

In order to increase the reliability and overall quality of our examples, we suggest to detach the `SpeechRecognized` and `SpeechRejected` event handlers and to stop the recognition activities once we close or unload the current window. The following code snippet provides a clean closure of the recognition process we may want to attach to the `Window.Closing` or `Window.Unload` event:

```
if (this.sensor != null)
{
    this.sensor.AudioSource.Stop();
    this.sensor.Stop();
    this.sensor = null;
}
if (null != this.speechEngine)
{
    this.speechEngine.SpeechRecognized -= SpeechRecognized;
    this.speechEngine.SpeechRecognitionRejected -= SpeechRejected;
    this.speechEngine.RecognizeAsyncStop();
}
```

We can certainly combine the Kinect sensor's speech recognition capabilities with the other strengths provided by the Kinect sensor itself.

For instance, we could enhance the example developed during the previous chapter and make a more fluid and natural way to select the image section.

The following grammar defined as an XML file represents the starting point to change the user experience of the example developed in *Chapter 3, Skeletal Tracking*.

```
<grammar version="1.0" xml:lang="en-US" root="rootRule" tag-
format="semantics/1.0-literals" xmlns="http://www.w3.org/2001/06/
grammar">
  <rule id="rootRule">
    <one-of>
      <item>
        <tag>SelectLEFT</tag>
        <one-of>
          <item> select left </item>
          <item> track left </item>
        </one-of>
      </item>
      <item>
        <tag>SelectRIGHT</tag>
        <one-of>
          <item> select right </item>
        </one-of>
      </item>
```

```
      <item>
        <tag>Deselect</tag>
        <one-of>
          <item> deselect </item>
        </one-of>
      </item>
    </one-of>
  </rule>
</grammar>
```

The following code snippet is the alternative or integration to the *Chapter 3* example's gesture commands:

```
private void SpeechRecognized(object sender, SpeechRecognizedEventArgs
e)
{    const double ConfidenceThreshold = 0.3;

     if (e.Result.Confidence >= ConfidenceThreshold)
     {
         switch (e.Result.Semantics.Value.ToString())
         {
             case "SelectLEFT":
                 if (!selected)
                 {
                     currentHand = JointType.HandLeft;
                     selected = true;
                 }
                 break;
             case "SelectRIGHT":
                 if (!selected)
                 {
                     currentHand = JointType.HandRight;
                     selected = true;
                 }
                 break;
             case "Deselect":
                 selected = false;
                 frameSelected = false;
                 startDrag = false;
                 break;

         }    }}
```

Tracking audio sources

The Kinect sensor captures sounds from every direction. Having said that, the audio capture hardware has a specified area (as an imaginary cone) from where it is able to capture audio signals the best. Audio waves that propagate through the length of the cone can be separated from the audio waves that travel across the cone.

Quite often we may find ourselves to be in noisy environments where different background noises and irrelevant sounds could corrupt the quality of our audio stream data analysis and/or the speech recognition. In this type of scenario we can improve the outcome of our application by pointing the cone in the direction of the audio that our application is most interested in capturing. In fact, we can improve the ability to capture and separate that audio source from other competing/distracting audio sources.

The Kinect sensor provides a few APIs for managing the audio source beam angle as well as the sound source angle. Using those APIs, we can set the direction of the imaginary cone and improve our ability to capture a specific audio source.

In the example we develop in this paragraph, we demonstrate how to use—at its best—the KinectAudioSource class and its members.

In our example we will combine the color stream data, the skeletal tracking, the audio stream data, and the speech recognition to provide a whole and articulated user experience.

Sound source angle

For simplicity, in the following code snippet we include only the part of codes that are relevant for the KinectAudioSource class. The complete example is included in this chapter's source code. The this.sensor instance is obtained as detailed in the previous chapters:

```
/// BeamAngleChanged event
this.sensor.AudioSource.BeamAngleChanged += this.
AudioSourceBeamChanged;

/// BeamAngleChanged event
this.sensor.AudioSource.SoundSourceAngleChanged += this.
AudioSourceSoundSourceAngleChanged;

///TUNING///

//Enables automatic gain control. The default value is false (no
automatic gain control).
this.sensor.AudioSource.AutomaticGainControlEnabled = true;
```

```
//Gets or sets the echo cancellation and suppression mode. The
default value is EchoCancellationMode.None.
    this.sensor.AudioSource.EchoCancellationMode =
EchoCancellationMode.CancellationAndSuppression;
By default noise suppression is enabled.
    //this.sensor.AudioSource.NoiseSuppression = true ;

    // Start streaming audio
    this.audioStream = this.sensor.AudioSource.Start();

    // Include the speech recognition initialization code snippet
    // defined in the previous paragraph HERE
}
```

The `KinectAudioSource` class supports three different modes for managing the `KinectAudioSource.BeamAngle` property. These modes are defined by the `KinectAudioSource.BeamAngleMode` property and detailed as per the following table:

Mode	Description
Automatic (default value)	Sets a beam angle and adapts it to the strongest audio source. It is recommended to use this setting for a low-volume loudspeaker and/or isotropic (with the same value when measured in different directions) background noise.
Adaptive	Sets the beam angle and adapts it to the strongest audio source. It is recommended to use this setting for a high-volume loudspeaker and/or higher noise levels.
Manual	Sets the beam angle to point in the direction of the audio source of interest.

Working on the `KinectAudioSource` properties such as `AutomaticGainControlEnabled`, `EchoCancellationMode`, `EchoCancellationSpeakerIndex`, and `NoiseSuppressionManual`, we can tune the audio source settings.

Handling the `KinectAudioSource.BeamAngleChanged` and `KinectAudioSource.SoundSourceAngleChanged` events, we can take actions for managing the direction of the imaginary cone and enforce our ability to capture a specific audio source.

Beam angle

Similar to the sound source angle, the beam angle is also defined in the x-z plane of the sensor perpendicular to the z-axis. The beam angle and the sound source angle are both updated continuously once the sensor has started streaming audio data (when the Start method is called).

In our example we are using the `AudioSourceBeamChanged` event handler to simply display — in the `tbBeamAngle` text block — the value of `BeamAngleChangedEventArgs.BeamAngle`.

As highlighted in the following code snippet, in the `AudioSourceSoundSourceAngleChanged` event handler we perform two main activities:

- Capture the level of confidence, which represents the confidence in the accuracy of the `SoundSourceAngle` property. This property is updated continuously after calling the `KinectAudioSource.Start` method.

- Project the `SoundSourceAngle` property to the color image captured by the color stream data. The computation applied for obtaining this projection (stored in `int audioSourcePosition`) is approximate and not meant to be scientific.

```
/// </summary>
/// <param name="sender">object sending the event.</param>
/// <param name="e">event arguments.</param>
private void AudioSourceSoundSourceAngleChanged(object sender,
SoundSourceAngleChangedEventArgs e)
{

    confidence = new SolidColorBrush();

    // convert the confidence value to a level of transparency.
    // Each value has a range of 0-255.
    confidence.Color = Color.FromArgb((byte)(255 * e.ConfidenceLevel),
255, 0, 0 );


    // we assume the -50 degrees match the left end of the image and
    // the +50 angle matches the right end. we linearly adjust all
    //the other values
    // in the interval

    audioSourcePosition = (int)((RenderWidth / 2) + ((RenderWidth / 2)
* e.Angle) / 50);

    tbSourceAngle.Text = string.Format(CultureInfo.CurrentCulture,
Properties.Resources.SourceAngle, e.Angle.ToString("0", CultureInfo.
CurrentCulture));
}
```

Within the `sensor_AllFramesReady` event handler, we draw vertical lines representing the following:

- The manual beam angle's projection in the image (aqua)

- The sound source angle's projection in the image (red with its transparency rated against the confidence value)

- The beam angle's projection in the image (green)

> Even though the red vertical line is an approximate projection of the source sound angle, it enables us to assert that there is an offset between the central point of the camera frame and the central point of the audio detection. We will notice those audios captured with a positive angle value are falling further on the right-hand side of the image, while audios captured with the opposite negative angle value are not symmetric on the left-hand side of the image. Hence our assertion: the two centers (the audio stream and the color stream ones) are misaligned.

```
void sensor_AllFramesReady(object sender, AllFramesReadyEventArgs e)
{

    //handle ColorImageFrame as in the previous chapters

    //handle the Skeleton stream data as in the previous chapter

    //draw the output
    using (DrawingContext dc = this.drawingGroup.Open())
    {
        // draw the color stream output
        dc.DrawImage(this.colorBitmap, new Rect(0.0, 0.0, RenderWidth,
RenderHeight));

        // Manage the BeamAngleMode.Manual Mode
        if (lstBeamAngleMode.SelectedIndex == 2)
        {
        // Manage the BeamAngleMode.Manual Mode
            if (selected)
            {
                foreach (Skeleton skeleton in this.skeletonData)
                {
                    if (skeleton != null)
                    {
                        Joint rightHand = skeleton.Joints[JointType.
HandRight];
```

```
                      if (rightHand != null && rightHand.
TrackingState == JointTrackingState.Tracked)
                            {

                                audioManualBeamPosition = (int)
SkeletonPointToScreen(rightHand.Position).X;

        // Track the ManualBeamAngle on the right hand position
                                this.sensor.AudioSource.ManualBeamAngle =
                                    Math.Atan(rightHand.Position.X /
rightHand.Position.Z) * (180 / Math.PI);

                                //tbManualBeamAngle.Text = string.
Format(CultureInfo.CurrentCulture, Properties.Resources.
ManualBeamAngle, this.sensor.AudioSource.ManualBeamAngle.ToString("0",
CultureInfo.CurrentCulture));
                                break;
                            }
                    } } }

        // Highlight with an Aqua line the ManualBeamAngle
            dc.DrawRectangle(Brushes.Aqua, null, new Rect( Math.Max(0,
audioManualBeamPosition - ClipBoundsThickness),
                            0, ClipBoundsThickness, RenderHeight));
        }

        // Highlight with Red line the SoundSourceAngle
        dc.DrawRectangle(confidence,
                null,
                new Rect(Math.Max(0, audioSourcePosition -
ClipBoundsThickness), 0, ClipBoundsThickness, RenderHeight));

        audioBeamPosition = (int)((RenderWidth / 2) + ((RenderWidth /
2) * this.sensor.AudioSource.BeamAngle) / 50);

        // Highlight with Green line the BeamAngle
        dc.DrawRectangle(Brushes.Green,
                null,
                new Rect(Math.Max(0, audioBeamPosition -
ClipBoundsThickness), 0, ClipBoundsThickness, RenderHeight));

        // define the limited area for rendering the visual outcome
        this.drawingGroup.ClipGeometry = new RectangleGeometry(new
Rect(0.0, 0.0, RenderWidth, RenderHeight));

    } }
```

Finally, we use the speech recognition for selecting and localizing the value of `ManualBeamAngle`. Voicing the `START` command, we start to drag the `ManualBeamAngle` value, activating the `selected = true` flag and moving our skeleton's right hand joint. Voicing the `OK` command, we confirm the `ManualBeamAngle` value, deactivating the `selected` flag:

```
/// <summary>
/// Handler for recognized speech events.
/// </summary>
/// <param name="sender">object sending the event.</param>
/// <param name="e">event arguments.</param>
private void SpeechRecognized(object sender, SpeechRecognizedEventArgs
e)
{
    // Speech utterance confidence below which we treat speech as if
    //it hadn't been heard
    const double ConfidenceThreshold = 0.3;

    if (e.Result.Confidence >= ConfidenceThreshold)
    {
        switch (e.Result.Semantics.Value.ToString())
        {   case "START":
                tbRecognizer.Text = "Recognized 'START' @ " +
e.Result.Confidence;
                selected = true;
                break;
            case "OK":
                tbRecognizer.Text = "Recognized 'OK' @ " + e.Result.
Confidence;
                selected = false;
                break;
            default:
                tbRecognizer.Text = "";
                break;
        }   }}
```

The `ManualBeamAngle` value is computed as per the following code snippet:

```
this.sensor.AudioSource.ManualBeamAngle = Math.Atan(rightHand.
Position.X / rightHand.Position.Z) * (180 / Math.PI);
```

Notice that the `SpeechRecognizedEventArgs.Result.Confidence` value decreases when we set the `ManualBeamAngle` value away from the angle between our mouth and the Kinect sensor. `SpeechRecognizedEventArgs.Result.Confidence` is near to 1 when the `ManualBeamAngle` value is equal to the angle between our mouth and the Kinect sensor.

In order to increase the reliability and overall quality of our examples, we suggest to detach the `AudioSourceBeamChanged` and `AudioSourceBeamChanged` event handlers and to stop the audio stream data activities once we close or unload the current window. The following code snippet provides a clean closure of the audio stream data processing we may want to attach to the `Window.Closing` or `Window.Unload` event:

```
void MainWindow_Unloaded(object sender, RoutedEventArgs e)
{
    if (this.sensor != null)
        {   this.sensor.AudioSource.BeamAngleChanged -=
this.AudioSourceBeamChanged;
            this.sensor.AudioSource.SoundSourceAngleChanged -= this.
AudioSourceSoundSourceAngleChanged;
            this.sensor.AudioSource.Stop();
            this.sensor.Stop();
            this.sensor = null;
}       }
```

The following image highlights the outcome of the example. On the left-hand side, we notice that the beam angle (green line) and the sound source (the mouth of the user approximated by the green line) are very close. On the right-hand side, we set manually the beam angle to a position (cyan line) that is far from the sound source (the mouth of the user). We can notice how this inducts the Kinect to incorrectly position the beam angle in the wrong direction (green line):

Projection of the sound source's angle, the beam angle, and the manual one

Summary

In this chapter we learned how to manage the Kinect sensor audio stream data and enhance the Kinect sensor's capabilities for speech recognition.

We have been working mainly using the `KinectAudioSource` class. This class manages the stream of either raw or modified audio from the microphone array. The audio stream can be modified to include a variety of algorithms to improve its quality, including noise suppression, automatic gain control, and acoustic echo cancellation.

First of all we introduced the concept of grammars for converting sounds in commands. Grammars are defined by XML files or programmatically. For increasing the quality of the speech recognition process, many times applications use specific prefixes to improve accuracy. While implementing a grammar, it is a good practice to define a speech command as a combination of application-specific keywords plus the actual command. This decreases the chance of treating random words as an actual speech command.

While working with the Kinect sensor, `Microsoft.Speech.Recognition.Grammar` is the class we need to use for defining and managing grammars processed by the `Microsoft.Speech.Recognition.SpeechRecognitionEngine` speech recognizer engine. We worked with the `SpeechRecognized` and `SpeechRecognizedRejected` events for managing the speech recognition process outcome defined as a result and level of confidence the speech recognition engine associates to the result itself.

The Kinect sensor captures sounds from every direction. Having said that, the sensor is also able to define an area of focus. Managing the `KinectAudioSource.BeamAngle` and `KinectAudioSource.BeamAngleMode` properties, we optimized the orientation of the imaginary cone in order to select the most relevant sound source we wanted to track.

Audio source settings can be tuned using the `KinectAudioSource` properties such as `AutomaticGainControlEnabled`, `EchoCancellationMode`, `EchoCancellationSpeakerIndex`, and `NoiseSuppressionManual`.

Managing the Kinect sensor accelerometer and its ability to move up and down the tilt motor can help us in optimizing the Kinect audio tracking results.

Speech recognition is a powerful capability handled by the Kinect sensor, which promotes the sensor to a first-class natural interface device. Combining its speech recognition capabilities together with its ability to track user's movements and gestures, Kinect enables us to build a true multimodal interface across all our examples.

In the last chapter we will focus on how the Kinect Studio can help us on debugging and analyzing the data stream provided by the Kinect sensor.

Kinect Studio and Audio Recording

In the previous chapters, we walked through the journey on how to manage and master all the data streamed out from the Kinect sensor, starting from managing the depth and color stream to implementing natural user interface enabled applications based on gestures and speech recognition.

While implementing the proposed examples, we have been standing up, walking in our room or office, and letting our colleagues or friends wonder what we were doing!

Of course, we would never like to discourage doing physical exercises and talking to our Kinect sensor, but having said so, there are in fact scenarios where we need to be close to the keyboard. For instance, when things go wrong and we cry for a passionate look through the source code flow (does that sound like a romantic way to explain *debugging*?). Moving back and forward from our keyboard limits our ability to spot issues. What about when we have to process the same stream of data over and over again, or, in a development team type of scenario, when we have to unit test the application in a repetitive manner?

In this appendix we will learn how we can save time coding and testing on Kinect enabled applications by:

- Recording all the video data coming into an application from a Kinect sensor with **Kinect Studio**
- Injecting the recorded video in an application allowing us to test our code without getting out of our chair over and over again
- Saving and playing back voice commands with a simple custom tool for enforcing quality on our application's speech recognition capabilities

The Kinect Studio is included in the Kinect for Windows Developer toolkit. You can download the same from `http://www.microsoft.com/en-us/download/details.aspx?id=27226`. You can install the Kinect for Windows Developer toolkit only after you have installed the Kinect for Windows SDK.

Kinect Studio – capturing Kinect data

Capturing data streamed out from the Kinect sensor with Kinect Studio is a simple and intuitive process. We need to run Kinect Studio and the **Connect to a Kinect App & Sensor** window, which will enable us to select the Kinect application from which we want to record the data RGB or depth streams.

Kinect Studio connection dialog

Thanks to this same window we can otherwise select the Kinect application to inject a stream we had saved earlier.

- Recording Kinect data is the operation we perform for creating, testing data, or testing the application on the fly.

- Injecting Kinect data is the operation we perform mainly for testing our application

We need to launch our application in advance to attach the same to Kinect Studio. In case we haven't launched our application yet, Kinect Studio will display an empty **Choose an app/sensor pair** list. Once we launch our application, we can select the **Refresh** button and the Kinect enabled application will eventually be listed.

By Kinect enabled applications we mean an application using the Kinect Sensor and where we have invoked the `KinectSensor. Start()` method.

Kinect Studio is able to track Kinect enabled applications wherever they are running in the debug or release mode. One additional requirement is that Kinect Studio and the Kinect enabled application should have the same level of authorization.

Once the Kinect enabled application is listed in the **Choose an app/sensor pair** list, we can select it and begin the test activities by clicking on the **Connect** button.

We are now ready to start our record activity against the color and depth data stream. We will go through the injection operation later.

We can select which stream we want to track, visualize, and eventually play back using the Kinect Studio main window.

 We can still record the depth stream even in cases where our application is not processing the RGB/depth stream. It is indeed sufficient to initialize the Kinect sensor and does not disable the IR stream in order to track the depth stream.

Usually, applications focused on speech recognition and/or audio beam positioning are not concerned about depth data. We should disable the IR stream in this scenario to improve the Kinect sensor's overall performance.

Let's now take a look at the scenario where we want to record a set of gestures to perform a repeatable test against our gesture recognition engine:

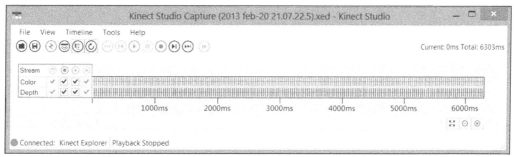

Main window

Select the **Record** graphical button (*Ctrl* + *R* using the keyboard). Kinect Studio starts to record the data stream from all the input sensors we have selected (color and/or depth). To stop recording the data stream we simply select the **Stop** graphical button (*Shift* + *F5* using the keyboard). Finally, using the **Play/Pause** graphical button (*Shift* + *F5* on the keyboard), we can play back the stream data we just recorded. All the data will be rendered in the **Depth Viewer**, **3D Viewer**, and **Color Viewer**, which provide a visual playback of the stream data. We will look more closely at the viewers window shortly.

We can save the stream data as .xed binary files. By selecting the **Save** graphical button (*Ctrl* + *S* using the keyboard), we make the test data persistent. The saved data can be injected in to the application and allows us to test the application without the need to stand once again in front of the Kinect.

 We recommend adopting the approach where we record a single gesture or key movement per single file. This will allow us to unit test the Kinect enabled applications.

Let's open the first recorded gesture. We can open a .xed file using the **Open** graphical button (*Ctrl* + *O* by the keyboard). We can fetch on a precise frame using the Kinect Studio timeline. Otherwise, we could select a given portion of the stream we want to reproduce. We can do this by selecting the starting point on the Kinect Studio timeline and dragging the mouse to the end point of the interval.

Using the **Play/Pause** graphical buttons, we can start playing back all the data contained in the recorded streams. The outcome is visualized in the different views as the following:

- In the **Depth Viewer** window we have the stream of the depth camera mapped with a color palette, which, in the SDK Version 1.6 of the Kinect for Windows, with colors between blue (for points close to the sensor) and red (for points far from the sensor). The black color is used for highlighting areas out of the range or not tracked.

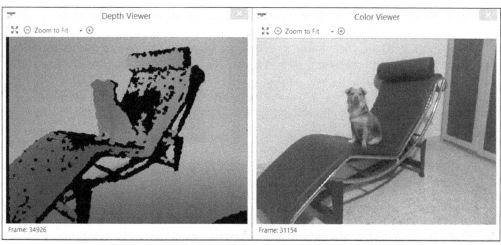

Depth Viewer window on left; Color Viewer window on right

- In the **Color Viewer** window we have the stream captured from the RGB camera.

- In the **3D Viewer** window we have the three-dimensional representation of the scene captured by the Kinect sensor. The viewer enables us to change the camera position and to have a different perspective of the very same scene.

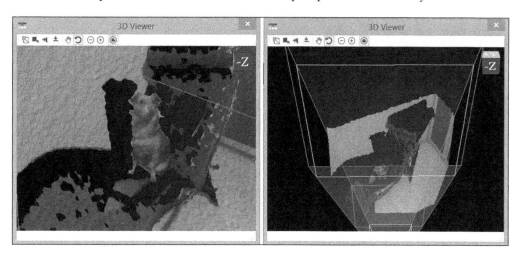

3D Viewer from two different perspectives

What makes Kinect Studio so vital is not only the fact that the recorded stream data is reproduced in the viewer, but indeed the fact that it can inject the same data in to our application. We can, for instance, record the stream data, notice a bug, fix our code, and then inject the very same stream data to ensure that the issue has been fixed. Kinect Studio is also very useful to ensure that the way we are rendering the stream data in our application is faithful. We can compare the graphical output of our application with the ones rendered by Kinect Studio and ensure that they are providing the same result, or rationalize the reason why they differ. For instance, in the following figure, we can understand that the color palette utilized in Kinect Studio for highlighting the depth points value is different from the one utilized in the application we developed in *Chapter 2, Starting with Image Streams*.

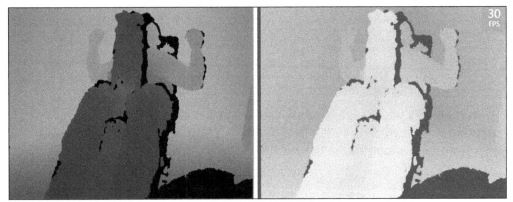

Depth Viewer on the left, Depth frame displayed inside app on the right

Audio stream data – recording and injecting

As stated previously, the Kinect Studio currently delivered by Microsoft does not support the tracking and injecting of the audio stream data.

In this appendix, we have attached a simple and primitive tool for recording the speech input and to submit it against the speech recognition engine and the grammar defined.

We encourage you to take the idea further and to realize a more complex and user-friendly Kinect Audio/Studio type of application.

The idea behind the tool is very simple. You can record your audio input as a .wav file and then inject it in to the speech recognition engine and debug/test the audio stream processing.

You may want to use a different `.wav` file and see how the speech engine recognition works against other people pronunciation or other environmental characteristics that differ from the one where you are currently testing your application. Have you ever thought of developing an application that is capturing commands from a song? Or what about building a chaos monkey (a small tool able to test the reliability of your application) type of test injecting a *no-sense* `.wav` file in to your application? How is the application reacting to that?

As you may remember, we enabled the speech recognition process in to *Chapter 4*, *Speech Recognition*, calling the key `SetInputToAudioStream` API of the `SpeechRecognitionEngine` class for processing the `AudioSource` streamed out from the `KinectSensor` (please refer to the following code snippet). This enabled our application to try recognizing all the speech inputs streamed in by the Kinect sensor:

```
speechEngine.SetInputToAudioStream(
    sensor.AudioSource.Start(),
    new SpeechAudioFormatInfo
    (EncodingFormat.Pcm, 16000, 16, 1, 32000, 2, null));
        speechEngine.RecognizeAsync(RecognizeMode.Multiple);
```

The `SpeechRecognitionEngine` class provides the `SetInputToWaveFile` method too, which enables us to receive input from a `.wav` file. So we can load the `.wav` file we recorded in advance with the following code:

```
speechEngine.SetInputToWaveFile("COMMAND_TO_TEST.WAV");
```

The speech recognition process will be the very same one we saw in the previous chapter. In order to save the audio captured by the Kinect sensors we can utilize the `Recorder` class to save the audio stream inside a `.wav` file format:

```
sealed class Recorder
{   static byte[] buffer = new byte[4096];
    static bool isRecording;
    public static bool IsRecording
    {   get { return isRecording; }
        set { isRecording = value; }
    }
```

The data format of a wave audio stream is defined by the `WAVEFORMATEX` structure:

```
struct WAVEFORMATEX
{   public ushort    wFormatTag;
    public ushort    nChannels;
    public uint      nSamplesPerSec;
    public uint      nAvgBytesPerSec;
    public ushort    nBlockAlign;
```

```
        public ushort    wBitsPerSample;
        public ushort    cbSize;
}
```

More details on a structure's members are explained in the Microsoft references at http://msdn.microsoft.com/en-us/library/ windows/hardware/ff538799(v=vs.85).aspx.

A complete list of WAVE_FORMAT_XXX formats (WAVE_FORMAT_PCM for one or two channel PCM data) can be found in the Mmreg.h header file.

With the WriteWavHeader method we create the header of the .wav file:

```
// Support method utilized by WriteWavHeader method
        static void WriteString(Stream stream, string s)
        {   byte[] bytes = Encoding.ASCII.GetBytes(s);
            stream.Write(bytes, 0, bytes.Length);
        }

        public static void WriteWavHeader(Stream stream, int
        dataLength)
        {   using (MemoryStream memStream = new MemoryStream(64))
            {   int cbFormat = 18;
                WAVEFORMATEX format = new WAVEFORMATEX()
                {   wFormatTag = 1,
                    nChannels = 1,
                    nSamplesPerSec = 16000,
                    nAvgBytesPerSec = 32000,
                    nBlockAlign = 2,
                    wBitsPerSample = 16,
                    cbSize = 0
                };

                using (var bw = new BinaryWriter(memStream))
                {   WriteString(memStream, "RIFF");
                    bw.Write(dataLength + cbFormat + 4);
                    WriteString(memStream, "WAVE");
                    WriteString(memStream, "fmt ");
```

```
                    bw.Write(cbFormat);
                    bw.Write(format.wFormatTag);
                    bw.Write(format.nChannels);
                    bw.Write(format.nSamplesPerSec);
                    bw.Write(format.nAvgBytesPerSec);
                    bw.Write(format.nBlockAlign);
                    bw.Write(format.wBitsPerSample);
                    bw.Write(format.cbSize);
                    WriteString(memStream, "data");
                    bw.Write(dataLength);
                    memStream.WriteTo(stream);
                }
            }}
```

The WriteWaveFile method converts the Kinect Audio source in the .wav file:

```
public static void WriteWavFile(KinectAudioSource sourceAudio,
FileStream fileStream)
        {   var size = 0;
            //Write header
            WriteWavHeader(fileStream, size);

            using (var audioStream = sourceAudio.Start())
            {   while (audioStream.Read(buffer, 0, buffer.Length) > 0
                && isRecording)
                {   fileStream.Write(buffer, 0, buffer.Length);
                    size += buffer.Length;
                }
                long prePosition = fileStream.Position;
                fileStream.Seek(0, SeekOrigin.Begin);
                WriteWavHeader(fileStream, size);
                fileStream.Seek(0, SeekOrigin.Begin);
                WriteWavHeader(fileStream, size);
                fileStream.Seek(prePosition, SeekOrigin.Begin);
                fileStream.Flush();
            }}
    }}
```

We recall the `Recorder` class inside our application simply by invoking the `RecordAudio` method:

```
private static object lockObject = new object();
private void RecordAudio()
        {
            lock (lockObject)
            {Recorder.IsRecording = true;
             using (var fileStream = new
                     FileStream("COMMAND.WAV", FileMode.Create))
             {
              Recorder.WriteWavFile(this.sensor.AudioSource,
              fileStream);
             }
            }
        }
```

To make our WPF application responsive to the user input and able to record the audio data streamed in by the Kinect sensor, we need to use background workers. The following code snippet highlights how to define the background worker and to invoke the `RecordAudio` method as the activity to implement when the background worker executes its work. The complete code source is provided in the code attached to this appendix:

```
private BackgroundWorker bgW =
new System.ComponentModel.BackgroundWorker();
...
this.bgW.RunWorkerCompleted += backgroundWorker1_RunWorkerCompleted;
this. bgW.DoWork += backgroundWorker1_DoWork;
...
void backgroundWorker1_DoWork(object sender, DoWorkEventArgs e)
{       RecordAudio();   }
...
Recorder.IsRecording = true;
if (!this.backgroundWorker1.IsBusy)
    {
this.backgroundWorker1.RunWorkerAsync();
    }
```

Summary

In this appendix we introduced Kinect Studio as a useful tool for testing our Kinect enabled application.

Kinect Studio can be installed with the Kinect for Windows Developer Toolkit. The Kinect for Windows SDK is the only software prerequisite for installing the Kinect for Windows Developer Toolkit. The Kinect for Windows Developer Toolkit is available as a free download at `http://www.microsoft.com/en-us/download/details.aspx?id=27226`.

Kinect Studio provides a simple interface to record and playback RGB and depth streams from a Kinect.

You can use the Kinect Studio recording capabilities for creating test data for the color and depth streams. Kinect Studio creates `.xed` binary files for all the color and depth data recorded during our testing sessions.

Thanks to the injection capability offered by the Kinect Studio we can test the video stream of our applications. This enables us to discover bugs and to apply solutions without the need to be away from our keyboard. As a matter of fact, even though the Kinect sensor is enabling our application to leverage a powerful multimodal interface, we are still depending on the keyboard for coding, analyzing performance, debugging our source code, and creating repeatable scenarios for testing.

Currently, Kinect Studio is not able to record and inject audio stream data. During this appendix we presented a simple and intuitive approach for testing the speech recognition process using a `.wav` file. The audio data streamed out by the Kinect sensor can be saved in a `.wav` file. Thanks to the `SpeechRecognitionEngine.SetInputToWaveFile()` method, we can exercise the speech recognition engine using the `.wav` file previously saved or any other `.wav` input.

Index

Thank you for buying
Kinect in Motion – Audio and Visual Tracking by Example

About Packt Publishing

Packt, pronounced 'packed', published its first book "*Mastering phpMyAdmin for Effective MySQL Management*" in April 2004 and subsequently continued to specialize in publishing highly focused books on specific technologies and solutions.

Our books and publications share the experiences of your fellow IT professionals in adapting and customizing today's systems, applications, and frameworks. Our solution based books give you the knowledge and power to customize the software and technologies you're using to get the job done. Packt books are more specific and less general than the IT books you have seen in the past. Our unique business model allows us to bring you more focused information, giving you more of what you need to know, and less of what you don't.

Packt is a modern, yet unique publishing company, which focuses on producing quality, cutting-edge books for communities of developers, administrators, and newbies alike. For more information, please visit our website: www.packtpub.com.

Writing for Packt

We welcome all inquiries from people who are interested in authoring. Book proposals should be sent to author@packtpub.com. If your book idea is still at an early stage and you would like to discuss it first before writing a formal book proposal, contact us; one of our commissioning editors will get in touch with you.

We're not just looking for published authors; if you have strong technical skills but no writing experience, our experienced editors can help you develop a writing career, or simply get some additional reward for your expertise.

Kinect for Windows SDK Programming Guide

ISBN: 978-1-84969-238-0   Paperback: 392 pages

Build motion-sensing applications with Microsoft's Kinect for Windows SDK quickly and easily

1. Building application using Kinect for Windows SDK

2. Covers the Kinect for Windows SDK v1.6

3. A practical step-by-step tutorial to make learning easy for a beginner

4. A detailed discussion of all the APIs involved and the explanations of their usage in detail

Cinder – Begin Creative Coding

ISBN: 978-1-84951-956-4   Paperback: 146 pages

A quick introduction into the world of creative coding with Cinder through basic tutorials and a couple of advanced examples

1. More power – Cinder is one of the most powerful creative coding engines out there and it will be hard to find a better one for your professional grade project

2. Do it fast – each section should not take longer than one hour to complete

3. We give you the tools and it is up to you what you do with them – we won't go into complicated algorithms, but rather give you the brushes and paints so you can paint the way you already know

Please check **www.PacktPub.com** for information on our titles

Cinema 4D R13 Cookbook

ISBN: 978-1-84969-186-4   Paperback: 514 pages

Elevate your art to the fourth dimension with Cinema 4D

1. Master all the important aspects of Cinema 4D

2. Learn how real-world knowledge of cameras and lighting translates onto a 3D canvas

3. Learn Advanced features like Mograph, Xpresso, and Dynamics.

4. Become an advanced Cinema 4D user with concise and effective recipes

Cinder Creative Coding Cookbook

ISBN: 978-1-84951-870-3   Paperback: 300 pages

Create compelling graphics, animation, and interaction with Kinect and Camera input using one of the most powerful C++ frameworks available

1. Learn powerful techniques for building creative applications using motion sensing and tracking

2. Create applications using multimedia content including video, audio, images, and text

3. Draw and animate in 2D and 3D using fast performance techniques

Please check **www.PacktPub.com** for information on our titles

www.ingramcontent.com/pod-product-compliance
Lightning Source LLC
LaVergne TN
LVHW080101070326
832902LV00014B/2357